Anaerobic Reactors

Biological Wastewater Treatment Series

The *Biological Wastewater Treatment* series is based on the book *Biological Wastewater Treatment in Warm Climate Regions* and on a highly acclaimed set of best selling textbooks. This international version is comprised by six textbooks giving a state-of-the-art presentation of the science and technology of biological wastewater treatment.

Titles in the *Biological Wastewater Treatment* series are:

Volume 1: *Wastewater Characteristics, Treatment and Disposal*
Volume 2: *Basic Principles of Wastewater Treatment*
Volume 3: *Waste Stabilisation Ponds*
Volume 4: *Anaerobic Reactors*
Volume 5: *Activated Sludge and Aerobic Biofilm Reactors*
Volume 6: *Sludge Treatment and Disposal*

Biological Wastewater Treatment Series

VOLUME FOUR

Anaerobic Reactors

Carlos Augusto de Lemos Chernicharo

Department of Sanitary and Environmental Engineering
Federal University of Minas Gerais, Brazil

Publishing
LONDON • NEW YORK

Published by IWA Publishing, Alliance House, 12 Caxton Street, London SW1H 0QS, UK

Telephone: +44 (0) 20 7654 5500; Fax: +44 (0) 20 7654 5555; Email: publications@iwap.co.uk
Website: **www.iwapublishing.com**

First published 2007
© 2007 IWA Publishing

Copy-edited and typeset by Aptara Inc., New Delhi, India

British Library Cataloguing in Publication Data
A CIP catalogue record for this book is available from the British Library

Library of Congress Cataloguing-in-Publication Data
A catalogue record for this book is available from the Library of Congress

ISBN: 1 84339 164 3
ISBN 13: 9781843391647

Contents

Contents

Preface

The present series of books has been produced based on the book *"Biological wastewater treatment in warm climate regions"*, written by the same authors and also published by IWA Publishing. The main idea behind this series is the sub-division of the original book into smaller books, which could be more easily purchased and used.

The implementation of wastewater treatment plants has been so far a challenge for most countries. Economical resources, political will, institutional strength and cultural background are important elements defining the trajectory of pollution control in many countries. Technological aspects are sometimes mentioned as being one of the reasons hindering further developments. However, as shown in this series of books, the vast array of available processes for the treatment of wastewater should be seen as an incentive, allowing the selection of the most appropriate solution in technical and economical terms for each community or catchment area. For almost all combinations of requirements in terms of effluent quality, land availability, construction and running costs, mechanisation level and operational simplicity there will be one or more suitable treatment processes.

Biological wastewater treatment is very much influenced by climate. Temperature plays a decisive role in some treatment processes, especially the natural-based and non-mechanised ones. Warm temperatures decrease land requirements, enhance conversion processes, increase removal efficiencies and make the utilisation of some treatment processes feasible. Some treatment processes, such as anaerobic reactors, may be utilised for diluted wastewater, such as domestic sewage, only in warm climate areas. Other processes, such as stabilisation ponds, may be applied in lower temperature regions, but occupying much larger areas and being subjected to a decrease in performance during winter. Other processes, such as activated sludge and aerobic biofilm reactors, are less dependent on temperature, as a result of the higher technological input and mechanisation level. The main purpose of this series of books is to present the technologies for urban wastewater treatment as applied to the specific condition of warm temperature, with the related implications in terms of design and operation. There is no strict definition for the range of temperatures that fall into this category, since the books always present how to correct parameters, rates and coefficients for different temperatures. In this sense, subtropical and even temperate climate are also indirectly covered, although most of the focus lies on the tropical climate.

Another important point is that most warm climate regions are situated in developing countries. Therefore, the books cast a special view on the reality of these countries, in which simple, economical and sustainable solutions are strongly demanded. All technologies presented in the books may be applied in developing countries, but of course they imply different requirements in terms of energy, equipment and operational skills. Whenever possible, simple solutions, approaches and technologies are presented and recommended.

Considering the difficulty in covering all different alternatives for wastewater collection, the books concentrate on off-site solutions, implying collection and transportation of the wastewater to treatment plants. No off-site solutions, such as latrines and septic tanks are analysed. Also, stronger focus is given to separate sewerage systems, although the basic concepts are still applicable to combined and mixed systems, especially under dry weather conditions. Furthermore, emphasis is given to urban wastewater, that is, mainly domestic sewage plus some additional small contribution from non-domestic sources, such as industries. Hence, the books are not directed specifically to industrial wastewater treatment, given the specificities of this type of effluent. Another specific view of the books is that they detail biological treatment processes. No physical-chemical wastewater treatment processes are covered, although some physical operations, such as sedimentation and aeration, are dealt with since they are an integral part of some biological treatment processes.

The books' proposal is to present in a balanced way theory and practice of wastewater treatment, so that a conscious selection, design and operation of the wastewater treatment process may be practised. Theory is considered essential for the understanding of the working principles of wastewater treatment. Practice is associated to the direct application of the concepts for conception, design and operation. In order to ensure the practical and didactic view of the series, 371 illustrations, 322 summary tables and 117 examples are included. All major wastewater treatment processes are covered by full and interlinked design examples which are built up throughout the series and the books, from the determination of the wastewater characteristics, the impact of the discharge into rivers and lakes, the design of several wastewater treatment processes and the design of the sludge treatment and disposal units.

The series is comprised by the following books, namely: *(1) Wastewater characteristics, treatment and disposal; (2) Basic principles of wastewater treatment; (3) Waste stabilisation ponds; (4) Anaerobic reactors; (5) Activated sludge and aerobic biofilm reactors; (6) Sludge treatment and disposal.*

Volume 1 (*Wastewater characteristics, treatment and disposal*) presents an integrated view of water quality and wastewater treatment, analysing wastewater characteristics (flow and major constituents), the impact of the discharge into receiving water bodies and a general overview of wastewater treatment and sludge treatment and disposal. Volume 1 is more introductory, and may be used as teaching material for undergraduate courses in Civil Engineering, Environmental Engineering, Environmental Sciences and related courses.

Volume 2 (*Basic principles of wastewater treatment*) is also introductory, but at a higher level of detailing. The core of this book is the unit operations and processes associated with biological wastewater treatment. The major topics covered are: microbiology and ecology of wastewater treatment; reaction kinetics and reactor hydraulics; conversion of organic and inorganic matter; sedimentation; aeration. Volume 2 may be used as part of postgraduate courses in Civil Engineering, Environmental Engineering, Environmental Sciences and related courses, either as part of disciplines on wastewater treatment or unit operations and processes.

Volumes 3 to 5 are the central part of the series, being structured according to the major wastewater treatment processes (*waste stabilisation ponds, anaerobic reactors, activated sludge and aerobic biofilm reactors*). In each volume, all major process technologies and variants are fully covered, including main concepts, working principles, expected removal efficiencies, design criteria, design examples, construction aspects and operational guidelines. Similarly to Volume 2, volumes 3 to 5 can be used in postgraduate courses in Civil Engineering, Environmental Engineering, Environmental Sciences and related courses.

Volume 6 (*Sludge treatment and disposal*) covers in detail sludge characteristics, production, treatment (thickening, dewatering, stabilisation, pathogens removal) and disposal (land application for agricultural purposes, sanitary landfills, landfarming and other methods). Environmental and public health issues are fully described. Possible academic uses for this part are same as those from volumes 3 to 5.

Besides being used as textbooks at academic institutions, it is believed that the series may be an important reference for practising professionals, such as engineers, biologists, chemists and environmental scientists, acting in consulting companies, water authorities and environmental agencies.

The present series is based on a consolidated, integrated and updated version of a series of six books written by the authors in Brazil, covering the topics presented in the current book, with the same concern for didactic approach and balance between theory and practice. The large success of the Brazilian books, used at most graduate and post-graduate courses at Brazilian universities, besides consulting companies and water and environmental agencies, was the driving force for the preparation of this international version.

In this version, the books aim at presenting consolidated technology based on worldwide experience available at the international literature. However, it should be recognised that a significant input comes from the Brazilian experience, considering the background and working practice of all authors. Brazil is a large country with many geographical, climatic, economical, social and cultural contrasts, reflecting well the reality encountered in many countries in the world. Besides, it should be mentioned that Brazil is currently one of the leading countries in the world on the application of anaerobic technology to domestic sewage treatment, and in the post-treatment of anaerobic effluents. Regarding this point, the authors would like to show their recognition for the Brazilian Research Programme on Basic Sanitation (PROSAB), which, through several years of intensive, applied, cooperative research has led to the consolidation of anaerobic treatment and aerobic/anaerobic post-treatment, which are currently widely applied in full-scale plants in Brazil. Consolidated results achieved by PROSAB are included in various parts of the book, representing invaluable and updated information applicable to warm climate regions.

Volumes 1 to 5 were written by the two main authors. Volume 6 counted with the invaluable participation of Cleverson Vitorio Andreoli and Fernando Fernandes, who acted as editors, and of several specialists, who acted as chapter authors: Aderlene Inês de Lara, Deize Dias Lopes, Dione Mari Morita, Eduardo Sabino Pegorini, Hilton Felício dos Santos, Marcelo Antonio Teixeira Pinto, Maurício Luduvice, Ricardo Franci Gonçalves, Sandra Márcia Cesário Pereira da Silva, Vanete Thomaz Soccol.

Many colleagues, students and professionals contributed with useful suggestions, reviews and incentives for the Brazilian books that were the seed for this international version. It would be impossible to list all of them here, but our heartfelt appreciation is acknowledged.

The authors would like to express their recognition for the support provided by the Department of Sanitary and Environmental Engineering at the Federal University of Minas Gerais, Brazil, at which the two authors work. The department provided institutional and financial support for this international version, which is in line with the university's view of expanding and disseminating knowledge to society.

Finally, the authors would like to show their appreciation to IWA Publishing, for their incentive and patience in following the development of this series throughout the years of hard work.

Marcos von Sperling
Carlos Augusto de Lemos Chernicharo

December 2006

The author

Carlos Augusto de Lemos Chernicharo
PhD in Environmental Engineering (Univ. Newcastle-upon-Tyne, UK). Associate professor at the Department of Sanitary and Environmental Engineering, Federal University of Minas Gerais, Brazil. Consultant to governmental and private companies in the field of wastewater treatment.
calemos@desa.ufmg.br

1

Introduction to anaerobic treatment

1.1 APPLICABILITY OF ANAEROBIC SYSTEMS

As a result of expanded knowledge, anaerobic sewage treatment systems, especially upflow anaerobic sludge blanket (UASB) reactors, have grown in maturity, occupying an outstanding position in several tropical countries in view of their favourable temperature conditions. Their acceptance changed from a phase of disbelief, which lasted until the early 1980s, to the current phase of widespread acceptance.

However, this great acceptance has frequently led to the development of projects and the implementation of treatment plants with serious conceptual problems. In this sense, the following chapters aim at providing information related to the principles, design and operation of anaerobic sewage treatment systems, with emphasis on upflow anaerobic sludge blanket reactors and anaerobic filters.

In principle, all organic compounds can be degraded by an anaerobic process, which is more efficient and economic when the waste is easily biodegradable.

Anaerobic digesters have been largely used in the treatment of solid wastes, including agricultural wastes, animal excrements, sludge from sewage treatment plants and urban wastes, and it is estimated that millions of anaerobic digesters have been built all over the world with this purpose. Anaerobic digestion has also been largely used in the treatment of effluents from agricultural, food and beverage industries, both in developed and developing countries, as shown in Table 1.1.

Also concerning the treatment of domestic sewage in warm-climate regions, a substantial increment has been verified in the use of anaerobic technology, notably by means of the UASB-type reactors. Naturally, in this case, the application of

Table 1.1. Main types of industries whose effluents can be treated by anaerobic process

Slaughterhouses and cold storage facilities	Alcohol production	Potato processing
Breweries	Starch production	Coffee processing
Leather factories	Yeast production	Fruit processing
Dairies	Soft drink production	Fish processing
Sugar refineries	Wine production	Vegetable processing

Source: GTZ/TBW (1997)

anaerobic technology depends much more on the temperature of the sewage, due to the low activity of anaerobic microorganisms at temperatures below 20 °C, and to the unfeasibility of heating the reactors. This is because domestic sewage is more diluted than industrial effluents, resulting in low volumetric production rates of methane gas, which makes its use as a source of heat energy uneconomical. Therefore, anaerobic treatment of domestic sewage becomes much more attractive for tropical- and subtropical-climate countries, which are mainly developing countries<indentry id=".a"></indentry>

1.2 POSITIVE ASPECTS

Several favourable characteristics of anaerobic systems, likely to be operated under high solids retention times and very low hydraulic detention times, provide them with great potential for application to the treatment of low-concentration wastewaters. They are also simple, low-cost technologies, with some advantages regarding operation and maintenance, as illustrated in Table 1.2.

Table 1.2. Advantages and disadvantages of the anaerobic processes

Advantages	Disadvantages
• Low production of solids, about 3 to 5 times lower than that in aerobic processes	• Anaerobic microorganisms are susceptible to inhibition by a large number of compounds
• Low energy consumption, usually associated with an influent pumping station, leading to very low operational costs	• Process start-up can be slow in the absence of adapted seed sludge
• Low land requirements	• Some form of post-treatment is usually necessary
• Low construction costs	• The biochemistry and microbiology of anaerobic digestion are complex, and still require further studies
• Production of methane, a highly calorific fuel gas	
• Possibility of preservation of the biomass, with no reactor feeding, for several months	• Possible generation of bad odours, although they are controllable
• Tolerance to high organic loads	• Possible generation of effluents with unpleasant aspect
• Application in small and large scale	• Unsatisfactory removal of nitrogen, phosphorus and pathogens
• Low nutrient consumption	

Source: Adapted from Chernicharo and Campos (1995); von Sperling (1995); Lettinga *et al.* (1996)

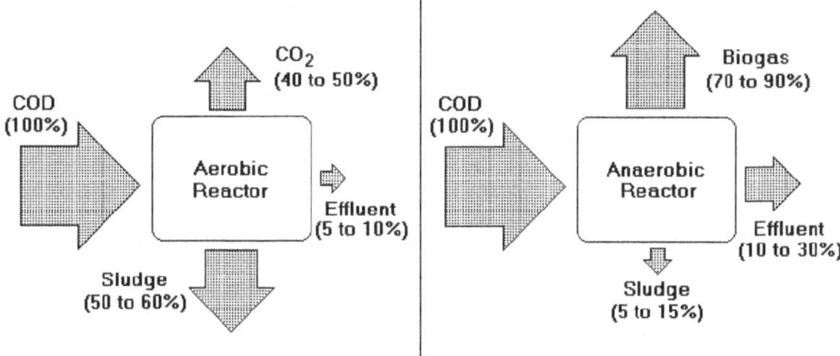

Figure 1.1. Biological conversion in aerobic and anaerobic systems

Figure 1.1 enables a clearer visualisation of some of the advantages of anaerobic digestion in relation to aerobic treatment, notably regarding the production of methane gas and the very low production of solids.

In *aerobic* systems, only about 40 to 50% of biological stabilisation occurs, with its consequent conversion into CO_2. A very large incorporation of organic matter as microbial biomass (about 50 to 60%) is verified, constituting the excess

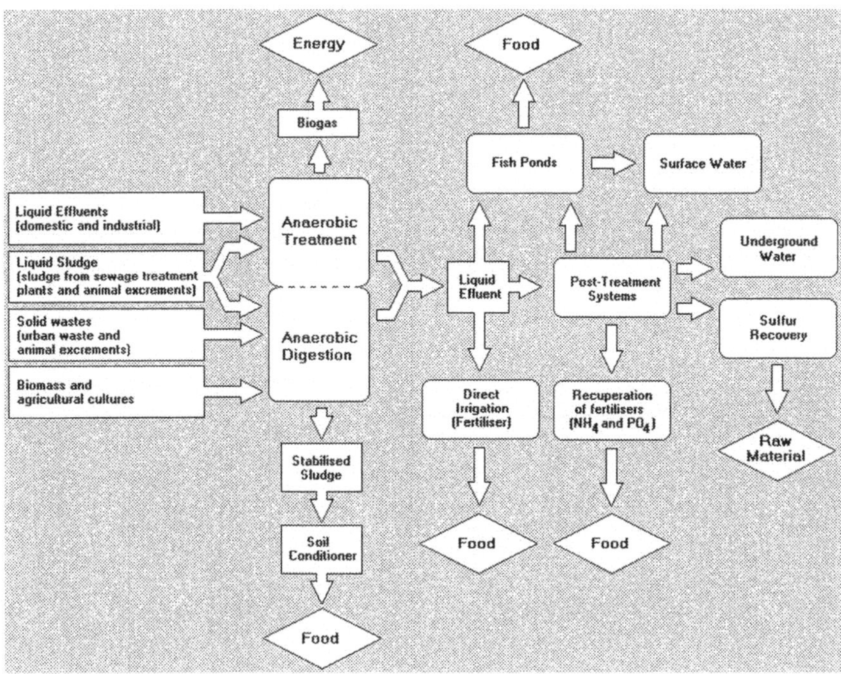

Figure 1.2. Anaerobic digestion as integrated technology for sewage treatment and by-product recovery (adapted from Lettinga, 1995)

sludge of the system. The organic material, not converted into carbon dioxide or into biomass, leaves the reactor as non-degraded material (5 to 10%).

In *anaerobic* systems, most of the biodegradable organic matter present in the waste is converted into biogas (about 70 to 90%), which is removed from the liquid phase and leaves the reactor in a gaseous form. Only a small portion of the organic material is converted into microbial biomass (about 5 to 15%), which then constitutes the excess sludge of the system. Besides the small amount produced, the excess sludge is usually more concentrated, with better dewatering characteristics. The material not converted into biogas or into biomass leaves the reactor as non-degraded material (10 to 30%).

Another interesting approach is made by Lettinga (1995), who emphasises the need for the implementation of integrated environmental protection systems that conciliate sewage treatment and the recovery and reuse of its by-products. The approach has a special appeal to developing countries, which present serious environmental problems, lack of resources and power and, frequently, insufficient food production. In this sense, anaerobic digestion becomes an excellent integrated alternative for sewage treatment and recovery of by-products, as illustrated in Figure 1.2.

2

Principles of anaerobic digestion

2.1 INTRODUCTION

Inorganic electron acceptors, such as SO_4^{2-} or CO_2, are used in the oxidation process of organic matter under anaerobic conditions. Methane formation does not occur in mediums where oxygen, nitrate or sulfate is readily available as electron acceptors. Methane production occurs in different natural environments, such as swamps, soil, river sediments, lakes and seas, as well as in the digestive organs of ruminant animals, where the redox potential is around -300 mV. It is estimated that anaerobic digestion with methane formation is responsible for the complete mineralisation of 5 to 10% of all the organic matter available on the Earth.

Anaerobic digestion represents an accurately balanced ecological system, where different populations of microorganisms present specialised functions, and the breakdown of organic compounds is usually considered a two-stage process. In the first stage, a group of facultative and anaerobic bacteria converts (by hydrolysis and fermentation) the complex organic compounds (carbohydrates, proteins and lipids) into simpler organic materials, mainly volatile fatty acids (VFA), as well as carbon dioxide and hydrogen gases.

In the second stage, the organic acids and hydrogen are converted into methane and carbon dioxide. This conversion is performed by a special group of microorganisms, named methanogens, which are strictly anaerobic prokaryotes. The methanogenic archaea depend on the substrate provided by the acid-forming microorganisms, consisting, therefore, in a syntrophic interaction.

The methanogens carry out two primordial functions in the anaerobic ecosystems: they produce an insoluble gas (methane) which enables the removal of organic carbon from the environment, and they also keep the H_2 partial pressure low enough to allow conditions in the medium for fermenting and acid-producing bacteria to produce more oxidised soluble products, such as acetic acid. Once the methanogens occupy the terminal position in the anaerobic environment during organic compound degradation, their inherent low growth rates usually represent a limiting factor in the digestion process as a whole.

2.2 MICROBIOLOGY OF ANAEROBIC DIGESTION

Anaerobic digestion can be considered an ecosystem where several groups of microorganisms work interactively in the conversion of complex organic matter into final products, such as methane, carbon dioxide, hydrogen sulfide, water and ammonia, besides new bacterial cells.

Although anaerobic digestion is generally considered a two-phase process, it can be subdivided into various metabolic pathways, with the participation of several microbial groups, each with a different physiological behaviour, as illustrated in Figure 2.1 and described in the following items.

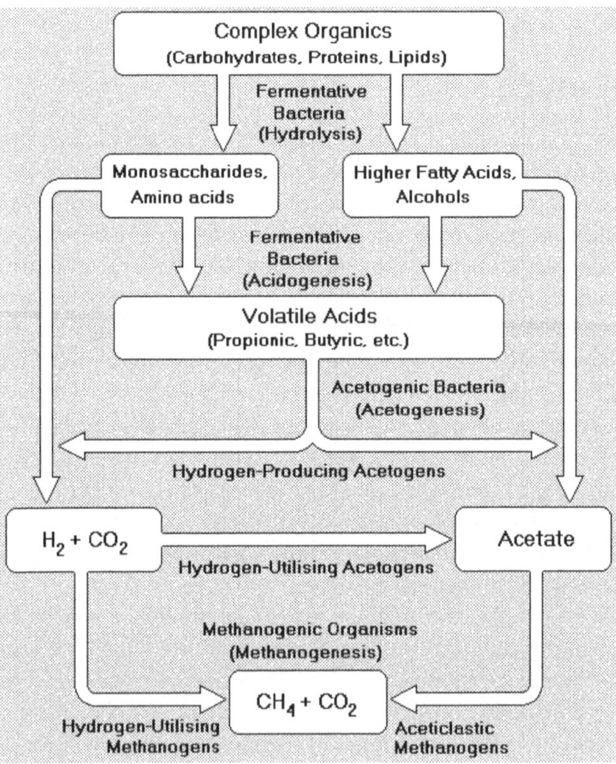

Figure 2.1.   Metabolic pathways and microbial groups involved in anaerobic digestion Adapted from: Lettinga *et al.* (1996)

(a) Hydrolysis and acidogenesis

Since the microorganisms are not capable of assimilating particulate organic matter, the first phase in the anaerobic digestion process consists in the hydrolysis of complex particulate material (polymers) into simpler dissolved materials (smaller molecules), which can penetrate through the cell membranes of the fermentative bacteria. Particulate materials are converted into dissolved materials by the action of exoenzymes excreted by the hydrolytic fermentative bacteria. The hydrolysis of polymers usually occurs slowly in anaerobic conditions, and several factors may affect the degree and rate at which the substrate is hydrolysed (Lettinga *et al.*, 1996):

- operational temperature of the reactor
- residence time of the substrate in the reactor
- substrate composition (e.g. lignin, carbohydrate, protein and fat contents)
- size of particles
- pH of the medium
- concentration of NH_4^+-N
- concentration of products from hydrolysis (e.g. volatile fatty acids)

The soluble products from the hydrolysis phase are metabolised inside the cells of the fermentative bacteria and are converted into several simpler compounds, which are then excreted by the cells. The compounds produced include volatile fatty acids, alcohols, lactic acid, carbon dioxide, hydrogen, ammonia and hydrogen sulfide, besides new bacterial cells.

Acidogenesis is carried out by a large and diverse group of fermentative bacteria. Usual species belong to the clostridia group, which comprises anaerobic species that form spores, able to survive in very adverse environments, and the family *Bacteroidaceaea*, organisms commonly found in digestive tracts, participating in the degradation of sugars and amino acids.

(b) Acetogenesis

Acetogenic bacteria are responsible for the oxidation of the products generated in the acidogenic phase into a substrate appropriate for the methanogenic microorganisms. In this way, acetogenic bacteria are part of an intermediate metabolic group that produces substrate for methanogenic microorganisms. The products generated by acetogenic bacteria are acetic acid, hydrogen and carbon dioxide.

During the formation of acetic and propionic acids, a large amount of hydrogen is formed, causing the pH in the aqueous medium to decrease. However, there are two ways by which hydrogen is consumed in the medium: (i) through the methanogenic microorganisms, that use hydrogen and carbon dioxide to produce methane; and (ii) through the formation of organic acids, such as propionic and butyric acids, which are formed through the reaction among hydrogen, carbon dioxide and acetic acid.

Among all the products metabolised by the acidogenic bacteria, only hydrogen and acetate can be directly used by the methanogenic microorganisms. However, at least 50% of the biodegradable COD are converted into propionic and butyric acids, which are later decomposed into acetic acid and hydrogen by the action of the acetogenic bacteria.

(c)   Methanogenesis

The final phase in the overall anaerobic degradation process of organic compounds into methane and carbon dioxide is performed by the methanogenic archaea. They use only a limited number of substrates, comprising acetic acid, hydrogen/carbon dioxide, formic acid, methanol, methylamines and carbon monoxide. In view of their affinity for substrate and extent of methane production, methanogenic microorganisms are divided into two main groups, one that forms methane from acetic acid or methanol, and the other that produces methane from hydrogen and carbon dioxide, as follows:

- acetate-using microorganisms (aceticlastic methanogens)
- hydrogen-using microorganisms (hydrogenotrophic methanogens)

Aceticlastic methanogens. Although only a few of the methanogenic species are capable of forming methane from acetate, these are usually the microorganisms prevailing in anaerobic digestion. They are responsible for about 60 to 70% of all the methane production, starting from the methyl group of the acetic acid. Two genera utilise acetate to produce methane: *Methanosarcina* prevails above 10^{-3} M acetate, while *Methanosaeta* prevails below this acetate level (Zinder, 1993). *Methanosaeta* may have lower yields and be more pH-sensitive, as compared to *Methanosarcina* (Schimidt and Ahring, 1996). *Methanosarcina* has a greater growth rate, while *Methanosaeta* needs a longer solids retention time, but can operate at lower acetate concentrations. The *Methanosaeta* genus is characterised by exclusive use of acetate, and having a higher affinity with it than the methanosarcinas. They are developed in the form of filaments, being largely important in the formation of the bacterial texture present in the granules. The organisms belonging to the *Methanosarcina* genus are developed in the form of coccus, which group together forming "packages". They are considered the most versatile among the methanogenic microorganisms, since they own species capable of using also hydrogen and methylamines (Soubes, 1994).

Hydrogenotrophic methanogens. Unlike the aceticlastic organisms, practically all the well-known methanogenic species are capable of producing methane from hydrogen and carbon dioxide. The genera more frequently isolated in anaerobic reactors are *Methanobacterium*, *Methanospirillum* and *Methanobrevibacter*. Both the aceticlastic and the hydrogenotrophic methanogenic microorganisms are very important in the maintenance of the course of anaerobic digestion, since they are responsible for the essential function of consuming the hydrogen produced in the previous phases. Consequently, the partial pressure of hydrogen in the medium is lowered, thus enabling the production reactions of the acidogenic and acetogenic bacteria (see Section 2.3.3).

(d) Sulfate reduction

In reactors treating wastewater containing sulfate or sulfite, these compounds can be used by sulfate-reducing bacteria (SRB) as acceptors of electrons released during the oxidation of organic materials (Lettinga *et al.*, 1996).

The metabolism of SRB is important in the anaerobic process, mostly because of their end product, hydrogen sulfide. SRB group species have in common the dissimilatory sulfate metabolism under strict anaerobiosis, and are considered a very versatile group of microorganisms, capable of using a wide range of substrate, including the whole chain of volatile fatty acids, several aromatic acids, hydrogen, methanol, ethanol, glycerol, sugars, amino acids and several phenol compounds.

Two major metabolic groups of SRB can be distinguished: (i) a group of species that is able to oxidise incompletely its substrates to acetate, like the genera *Desulfobulbus* sp. and *Desulfomonas* sp., and most of the species of the genera *Desulfotomaculum* and *Desulfovibrio* belong to this group; and (ii) a group which is able to oxidise its organic substrates, including acetate, to carbon dioxide. The genera *Desulfobacter*, *Desulfococcus*, *Desulfosarcina*, *Desulfobacterium* and *Desulfonema* belong to this group.

In the absence of sulfate, the anaerobic digestion process occurs according to the metabolic sequences presented in Figure 2.1. With the presence of sulfate in the wastewater, many of the intermediate compounds formed by means of the metabolic routes identified in Figure 2.1 start to be used by the SRB, causing a change in the metabolic routes in the anaerobic digester (see Figure 2.2). Hence, the SRB start to compete with the fermentative, acetogenic and methanogenic microorganisms for the substrate available, resulting in a decrease in the production of methane from a given amount or organic material present in the influent. The importance of this bacterial competition is greater when the relative concentration of SO_4^{2-} is increased in relation to the COD concentration (see Section 2.3.7).

2.3 BIOCHEMISTRY OF ANAEROBIC DIGESTION

2.3.1 Preliminaries

Anaerobic digestion of organic compounds comprises several types of methanogenic and acidogenic microorganisms, and the establishment of an ecological balance among the types and species of anaerobic microorganisms is of fundamental importance to the efficiency of the treatment system. The VFA parameter is frequently used for the evaluation of this ecological balance.

The volatile fatty acids are formed, as intermediate products, during the degradation of carbohydrates, proteins and lipids. The most important components resulting from the biochemical decomposition of the organic matter are the short-chain volatile acids, such as formic, acetic, propionic, butyric and, in smaller amounts, valeric and isovaleric acids. These low-molecular-weight fatty acids are named volatile acids because they can be distilled at atmospheric pressure. The volatile acids represent intermediate compounds, from which most of the methane is produced, through conversion by the methanogenic microorganisms.

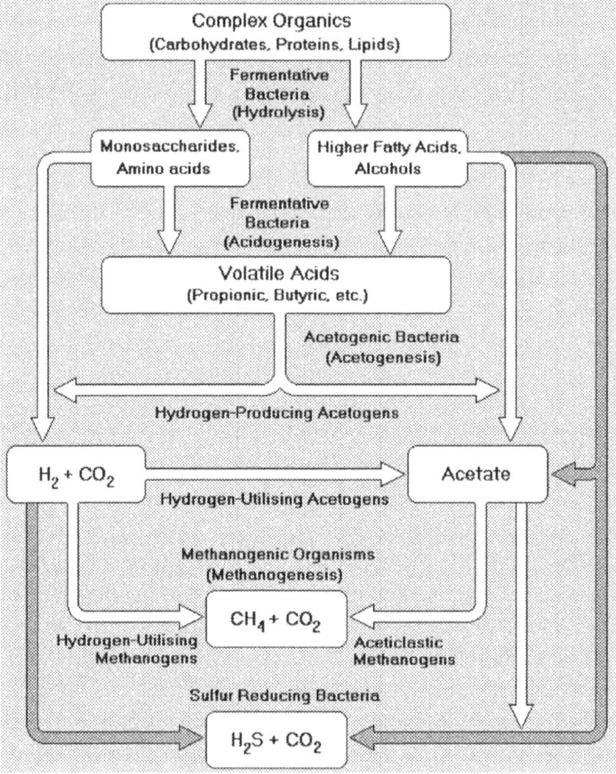

Figure 2.2.   Metabolic pathways and microbial groups involved in anaerobic digestion (with sulfate reduction). Source: Adapted from Lettinga *et al.* (1996)

When a population of methanogenic microorganisms is present in a sufficient amount, and the environmental conditions inside the treatment system are favourable, they use the intermediate acids as quickly as they are formed. Consequently, the acids do not accumulate beyond the neutralising capacity of the alkalinity naturally present in the medium, the pH remains in a range favourable for the methanogenic organisms and the anaerobic system is balanced. However, if the methanogenic organisms are not present in sufficient amount, or if they are exposed to unfavourable environmental conditions, they will not be capable of using the volatile acids at the same rate at which they are produced by the acidogenic bacteria, resulting in an accumulation of acids in the system. In these conditions, the alkalinity is quickly consumed, and the non-neutralised free acids cause the pH to drop. When that occurs the reactor is referred to by operators as 'sour' (because of its odour).

An identification of the individual acids present in a reactor with unbalanced bacterial populations can indicate which types of methanogenic microorganisms are not fulfilling their role in the treatment.

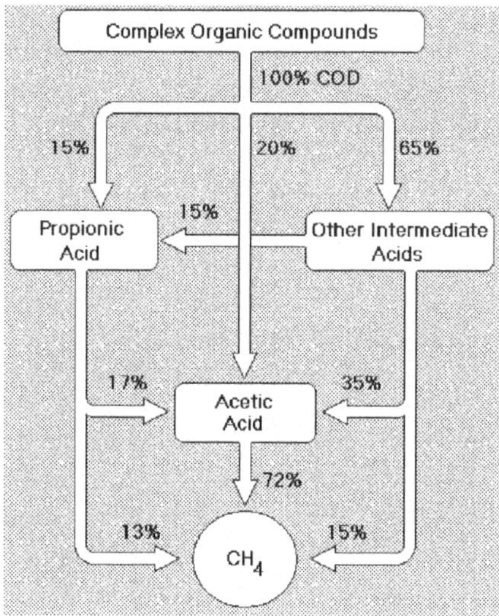

Figure 2.3.   Methane formation routes from the fermentation of complex substrates (adapted from McCarty, 1964)

2.3.2 Intermediate volatile acids

The most important intermediate volatile acids, precursors of methane formation, are the acetic and propionic acids. Some of the various metabolic steps involved in the degradation of a complex substrate, such as the excess sludge from domestic sewage treatment plants, are shown in Figure 2.3. The percentages shown are based on COD conversion, valid only for the formation of methane from complex substrates, such as sludges from sewage treatment plants or others of similar composition.

For the complete fermentation of complex compounds into methane, each group of microorganisms has a specific function. Even if the contribution to the process as a whole is small, it is nevertheless necessary for the formation of the final product. Propionic acid results mainly from the fermentation of the carbohydrates and proteins present, and about 30% of the organic compounds are converted into this acid before they can be finally converted into methane. Acetic acid is the most abundant intermediate acid, formed from all the organic compounds. Concerning the degradation of complex substrates, such as sludge from sewage treatment plants, acetic acid is precursor of about 72% of the methane formed and, together with propionic acid, of about 85% of the total methane production. A large part of the remaining 15% results from the degradation of other acids, such as formic and butyric acids.

2.3.3 Thermodynamic aspects

Some of the conversion reactions of the products from fermentative bacteria into acetate, hydrogen and carbon dioxide are illustrated in Table 2.1. The last column

Table 2.1. Some important oxi-reduction reactions in anaerobic digestion

Nr	Oxidation reactions (electron donors)		ΔG_o (kJ/mole)
1	Propionate $\Rightarrow$ acetate	$CH_3CH_2COO^- + 3H_2O$ $\Rightarrow CH_3COO^- + HCO_3^- + H^+ + 3H_2$	$+76.1$
2	Butyrate $\Rightarrow$ acetate	$CH_3CH_2CH_2COO^- + 2H_2O$ $\Rightarrow 2CH_3COO^- + H^+ + 2H_2$	$+48.1$
3	Ethanol $\Rightarrow$ acetate	$CH_3CH_2OH + H_2O$ $\Rightarrow CH_3COO^- + H^+ + 2H_2$	$+9.6$
4	Lactate $\Rightarrow$ acetate	$CH_3CHOHCOO^- + 2H_2O$ $\Rightarrow CH_3COO^- + HCO_3^- + H^+ + 2H_2$	-4.2
	Reduction reactions (electron acceptors)		
5	Bicarbonate $\Rightarrow$ acetate	$2HCO_3^- + 4H_2 + H^+$ $\Rightarrow CH_3COO^- + 4H_2O$	-104.6
6	Bicarbonate $\Rightarrow$ methane	$HCO_3^- + 4H_2 + H^+ \Rightarrow CH_4 + 3H_2O$	-135.6
7	Sulfate $\Rightarrow$ sulfide	$SO_4^{2-} + 4H_2 + H^+ \Rightarrow HS^- + 4H_2O$	$-151,9$

Source: Adapted from Foresti (1994) and Lettinga *et al.* (1996)

of the table shows the variation of standard free energy (pH equal to 7 and pressure of 1 atm), considering a temperature of 25 °C and the liquid being pure water. All the compounds present in the solution show a 1 mole/kg activity.

In accordance with the examples presented in Table 2.1, it can be clearly noticed that propionate, butyrate and ethanol (reactions 1, 2 and 3) are not degraded under the assumed standard conditions, as the thermodynamic aspects are unfavourable ($\Delta G_o > 0$). However, should the hydrogen concentration be low, the reactions can move to the right (product side). In practice, this is achieved by the continuous removal of H_2 from the medium, by means of electron acceptor reactions (e.g. reactions 5, 6 and 7). In a methanogenic digester operating in an appropriate manner, the partial H_2 pressure does not exceed 10^{-4} atm, and usually this pressure is close to 10^{-6} atm. Under these conditions of low partial hydrogen pressure, propionate, butyrate and ethanol start to degrade and release free energy to the medium. These low partial pressures can only be maintained if the hydrogen formed is quickly and effectively removed by the hydrogen-consuming microorganisms (Lettinga *et al.*, 1996).

2.3.4 Methane formation

Although the individual pathways involved in methane formation are not completely established yet, substantial progress in their understanding has been made in the past decades. Some methanogenic species are capable of using just hydrogen and carbon dioxide for their growth and methane formation, while others are capable of using formic acid, which is previously converted into hydrogen and carbon dioxide. At least two *Methanosarcina* species are capable of forming methane from methanol or acetic acid.

There are two basic mechanisms for methane formation: (i) cleavage of acetic acid and (ii) reduction of carbon dioxide. These mechanisms can be described as follows. In the absence of hydrogen, cleavage of acetic acid leads to the formation of methane and carbon dioxide. The methyl group of the acetic acid is reduced to methane, while the carboxylic group is oxidised to carbon dioxide:

$$C^*H_3COOH \Rightarrow C^*H_4 + CO_2 \qquad (2.1)$$

Microbial group involved: aceticlastic methanogenic organisms

When hydrogen is available, most of the remaining methane is formed from the reduction of carbon dioxide. CO_2 acts as an acceptor of the hydrogen atoms removed from the organic compounds by the enzymes. Since carbon dioxide is always present in excess in an anaerobic reactor, its reduction to methane is not the limiting factor in the process. The methane formation from the reduction of the carbon dioxide is shown below:

$$CO_2 + 4H_2 \Rightarrow CH_4 + 2H_2O \qquad (2.2)$$

Microbial group involved: hydrogenotrophic methanogenic organisms

The overall composition of the biogas produced during anaerobic digestion varies according to the environmental conditions prevailing in the reactor. The composition changes quickly during the initial start-up of the system and also when the digestion process is inhibited. For reactors operating in a stable manner, the composition of the biogas produced is reasonably uniform. However, the carbon dioxide/methane ratio can vary substantially, depending on the characteristics of the organic compound to be degraded. In the anaerobic treatment of domestic sewage, typical methane and carbon dioxide fractions present in the biogas are 70 to 80% and 20 to 30%, respectively.

The methane produced in anaerobic digestion processes is quickly separated from the liquid phase due to its low solubility in water. This results in a high degree of degradation of the liquid wastes, once this gas leaves the reactor to the gaseous phase. On the other hand, carbon dioxide is much more soluble in water than methane, and leaves the reactor partly as gas and partly dissolved in the liquid effluent.

2.3.5 Wastewater characteristics and COD balance

Although practical experience in the anaerobic treatment of liquid effluents is still recent, the potential application of the process can be evaluated from the knowledge of a few chemical characteristics of the waste to be treated. A preliminary evaluation of these characteristics will help choose the most suitable treatment process, allowing an estimation of biological solids production, nutrient requirements, methane production, etc.

Wastewater concentration in terms of biodegradable solids is of fundamental importance, and it can be reasonably estimated from the BOD and COD tests. Another important factor to be considered is the relative concentration of carbohydrates, proteins and lipids, in addition to other important chemical characteristics of the anaerobic biological treatment, especially pH, alkalinity, inorganic nutrients, temperature and the occasional presence of potentially toxic compounds.

(a) COD balance

Hulshoff Pol (1995) presented important and detailed considerations on the COD balance throughout the anaerobic degradation process. According to the author, the compounds present in the wastewater can be classified as of easy, difficult, or impossible degradation. Easily degradable compounds are those that are readily fermented by any type of anaerobic biomass (adapted or not to the waste type). The compounds of difficult degradation, named *complex substrates*, are not fermented by anaerobic microorganisms prior to their adaptation to the substrate. The period of adaptation to the substrate reflects the growth time of specialised microorganisms that can ferment the complex substrate. Lastly, certain organic compounds, known as *inert organic compounds*, are absolutely impossible to biologically degrade in anaerobic environments.

Biodegradable COD. Biodegradable COD (COD_{bd}) is a means of expressing the sewage treatability, which is defined as the total COD (COD_{tot}) portion present in the waste that can be biologically degraded in anaerobic conditions. The sewage biodegradability percentage is given by:

$$\%COD_{bd} = \frac{COD_{bd}}{COD_{tot}} \times 100 \tag{2.3}$$

where:
$\%COD_{bd}$ = percentage of biodegradable COD (%)
COD_{bd} = concentration of biodegradable COD (mg/L)
COD_{tot} = concentration of total COD (mg/L)

Acidifiable COD. In an anaerobic reactor, the raw sewage provides the fermentative bacteria with non-acidified biodegradable substrate (COD_{bd}). This substrate is consumed by the fermentative microorganisms and converted into cells (COD_{cel}), hydrogen and volatile fatty acids. It is assumed that not all the COD_{bd} will be available for the methanogenic microorganisms, once part of it is converted into new bacterial cells. The COD_{bd} fraction that will be truly available for the methanogenic microorganisms is named acidified COD (COD_{acid}). Thus, the amount of influent biodegradable COD (COD_{inf}) that can be acidified is the sum of the fractions converted into VFA and methane (CH_4). The sewage acidification percentage can then be expressed as follows:

$$\%COD_{acid} = \frac{COD_{CH_4} + COD_{VFA}}{COD_{inf}} \times 100 \tag{2.4}$$

where:
$\%COD_{acid}$ = percentage of acidified COD (%)
COD_{inf} = biodegradable COD contained in the influent (mg/L)

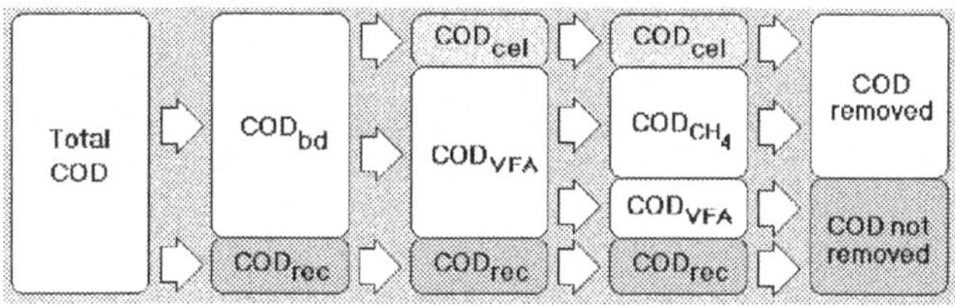

Figure 2.4. Diagram of the COD balance throughout the anaerobic degradation process

COD_{CH_4} = fraction of influent COD converted into methane (mg/L)
COD_{VFA} = fraction of COD still present as volatile fatty acids in the effluent (mg/L)

Recalcitrant COD. The recalcitrant COD (also named biologically resistant COD (COD_{rec})) refers to the portion of organic substrate that cannot be degraded by the fermentative microorganisms. The COD_{rec} is due to the complex substrate subjected to treatment in anaerobic reactors containing biomass not yet adapted to the complex substrate, or to the substrate considered biologically inert. Hence, the COD_{rec} is not fermented, and left biologically unaffected in the treated effluent. Figure 2.4 shows the COD balance throughout the anaerobic degradation process.

Soluble and particulate COD. Most of the compounds present in the raw sewage are not originally soluble and, added to the cells produced during the COD_{bd} degradation process, they form the portion of insoluble or particulate COD (COD_{part}). The COD solubility is usually known by means of laboratory analyses, and it may be presented in three types:

- **Filtered COD (COD_{filt})**. It is due to the presence of dissolved organic compounds in a sewage sample. The COD_{filt} is determined by using the portion of sample that passes through a paper filter of known pore size (1.5 µm). Alternatively to filtration, the sample can be centrifuged (5,000 rpm for 5 minutes), and the COD_{filt} from the supernatant liquid can be determined.
- **Particulate COD (COD_{part})**. It is due to the presence of suspended organic solids contained in a sewage sample. The COD_{part} is obtained as the difference between the total COD (sample neither filtered nor centrifuged) and the COD_{filt}, that is, the particulate COD is due to the solids which do not pass through the filter paper or that remain at the bottom of the recipients after the centrifugation stage.
- **Soluble COD (COD_{sol}):** The COD_{filt} of a sewage sample includes both the portion due to the dissolved particles (totally soluble) and the portion due to the presence of colloidal particles. The latter, responsible for the turbidity, is not removed by the conventional filtration or centrifugation

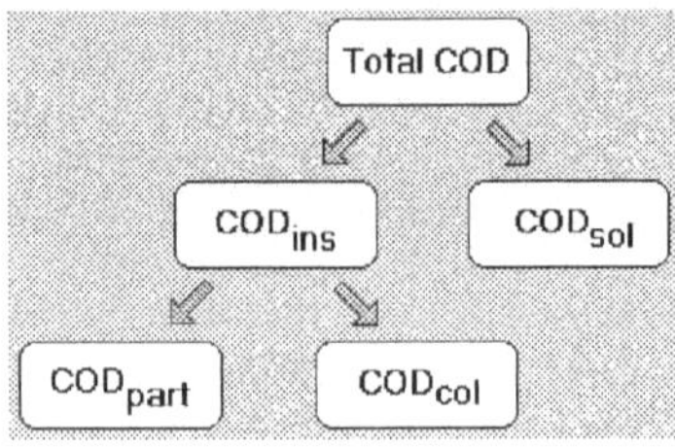

Figure 2.5.   Classification of the sewage COD according to solubility

methods. This way, the real COD_{sol} consists of the portion of COD_{filt} that passes through a membrane filter.

Based on these considerations, the following relations can be established (see also Figure 2.5):

$$COD_{tot} = COD_{part} + COD_{col} + COD_{sol} \qquad (2.5)$$

$$COD_{ins} = COD_{part} + COD_{col} \qquad (2.6)$$

$$COD_{fil} = COD_{col} + COD_{sol} \qquad (2.7)$$

Hydrolysable COD. Sewage usually contains organic polymers that need to be converted into simpler substrates (monomers) before being fermented. These organic compounds constitute the portion of hydrolysable COD, and the percentage of effectively hydrolysed insoluble COD is given by:

$$\%COD_{hid} = \frac{COD_{sol} + COD_{cel} + COD_{CH_4}}{COD_{ins}} \times 100 \qquad (2.8)$$

where:

$\quad \%COD_{hid}$ = percentage of hydrolysed COD (%)

$\quad COD_{sol}$ = fraction of soluble COD (including the volatile fatty acids) (mg/L)

$\quad COD_{cel}$ = fraction of COD converted into new fermentative bacteria cells (mg/L)

$\quad COD_{CH_4}$ = fraction of COD converted into methane (mg/L)

$\quad COD_{ins}$ = fraction of insoluble COD (particulate substrate) (mg/L)

(b)   COD removal

The removal of COD in an anaerobic reactor may occur in two ways:

Biological COD removal

The elimination of soluble COD in the system refers to the difference between the influent COD and the effluent COD, and the COD removal percentage is expressed by:

$$\%COD_{remov} = \frac{COD_{inf} - COD_{eff}}{COD_{inf}} \times 100 \tag{2.9}$$

where:

$\%COD_{remov}$ = percentage of COD removed (%)

COD_{inf} = concentration of influent COD (mg/L)

COD_{eff} = concentration of effluent COD (mg/L)

Considering that the total COD of the effluent comprises the particulate COD due to the microorganism cells, there is generally a greater significance in working with the filtered COD of the effluent, which enables the identification of the COD fraction used for cellular growth as follows:

$$\%COD_{cel} = \frac{\%removalCOD_{fil} - \%COD_{CH_4}}{\%removalCOD_{fil} + \%COD_{VFA}} \times 100 \tag{2.10}$$

where:

$\%COD_{cel}$ = percentage of COD converted into new cells (%)

$\%removal\ COD_{fil}$ = percentage of removal of filtered COD related to the influent soluble COD (%)

$\%COD_{CH_4}$ = percentage of COD converted into methane (%)

$\%COD_{VFA}$ = percentage of influent COD still present as VFA in the effluent (%)

When the influent COD is already acidified, that is, already converted into volatile fatty acids, the elimination percentage of filtered COD is approximately equal to the percentage of COD converted into methane, since the yield coefficient of the methanogenic microorganisms is very low.

The preceding considerations refer to the biological removal of soluble COD. The evaluation of the biological removal of insoluble COD (particulate) is more difficult, since the portion of particulate COD non-hydrolysed and non-degraded in the system cannot be distinguished from the bacterial cells present in the effluent.

Non-biological removal of COD

Non-biological mechanisms of removal of soluble COD usually occur in biological sewage treatment systems, through their incorporation either in the sludge or in the particulate fraction lost with the effluent. In these cases, the percentage of removal of filtered COD will include a portion of COD eliminated by non-biological insolubility. Two main mechanisms contribute to that: precipitation and adsorption:

Precipitation usually results from changes in the pH or from the addition of calcium-based alkaline compounds, for pH control. The precipitates can settle, and then be incorporated into the sludge or be taken out from the system together with the effluent COD.

Adsorption consists in a reaction where the soluble COD is adsorbed on the surface of the biomass particles present in the system. The most important example in practice is the fat adsorption on the bacterial sludge.

In addition, a portion of insoluble COD (particulate) can be removed by non-biological mechanisms, by means of its retention in the sludge. Such retention occurs because the sludge bed can act as a "filter" or because the particulate material can have good settleability.

In the specific case of UASB reactors (see Chapter 5), or of any other anaerobic system that depends on the immobilisation of active biomass, the accumulation of insoluble COD on the sludge bed can be harmful to the process. This accumulation causes the formation of non-bacterial sludge which, if in excess, can cause dilution of the population of methanogenic microorganisms in the sludge, thus reducing the methanogenic activity.

2.3.6 Wastewater degradation and methane production

As described in Section 2.2, anaerobic digestion can be considered a two-phase process. In the first phase, a diversity of fermentative bacteria initially converts the complex organic compounds into soluble compounds and, at last, into short-chain volatile fatty acids. In the second phase, the methanogenic microorganisms use the products fermented in the first phase and convert them into methane. If hydrogen is not produced in the first phase, the fermentation stage results in an insignificant reduction of COD, once all the electrons released in the oxidation process of the organic compounds are transferred to organic acceptors, which remain in the medium. Hence, even though the fermentation stage enables the conversion of part of the energy source into carbon dioxide and of part of the organic matter into new cells, it is considered an inadequate process for both the return of organic carbon to the atmosphere and its removal from the wastewater. However, when hydrogen is formed, it represents a gaseous product that escapes from the medium, causing, therefore, a reduction in the energy content of the wastewater.

Many of the acids and alcohols produced in the initial fermentation phase are converted into a highly insoluble gas, methane, that escapes from the medium, thus favouring the main mechanism for recycling of the organic carbon under anaerobic conditions. Except for the losses caused by microbial inefficiency, almost all the energy removed from the system is recovered in the form of methane gas. However, the formation of methane does not complete the carbon cycle, unless it is oxidised into carbon dioxide, either biologically or by combustion, to become available for recycling by photosynthesis.

(a)   Estimation of methane production considering the chemical composition of the waste

Knowing the chemical composition of the wastewater enables an estimation of the amount of methane to be produced and, consequently, of the amount of degraded organic matter. The Buswell stoichiometric equation is used to estimate the production of methane from a given chemical composition of the wastewater:

$$C_n H_a O_b N_d + \left(n - \frac{a}{4} - \frac{b}{2} + \frac{3d}{4}\right) H_2O$$
$$\Rightarrow CH_4 + \left(\frac{n}{2} - \frac{a}{8} + \frac{b}{4} + \frac{3d}{8}\right) CO_2 + (d)\,NH_3 \tag{2.11}$$

In this equation, $C_n H_a O_b N_d$ represents the chemical formula of the biodegradable organic compound subjected to the anaerobic degradation process, and the production of methane considered herein is the maximum stoichiometrically possible. Neither the use of substrate nor other routes of conversion of organic matter are taken into consideration for the production of bacterial biomass.

In the presence of oxygen (less probable) or of specific inorganic donors (such as nitrate, sulfate or sulfite), the production of methane will decrease, according to the following equations (Lettinga *et al.*, 1996):

$$10H + 2H^+ + 2NO_3^- \Leftrightarrow N_2 + 6H_2O \tag{2.12}$$

(considering the presence of nitrate in the wastewater)

$$8H + SO_4^{2-} \Leftrightarrow H_2S + 2H_2O + 2OH^- \tag{2.13}$$

(considering the presence of sulfate in the wastewater)

Equation 2.13 shows that the reduced sulfate in an anaerobic reactor leads to the formation of H_2S, a gas that dissolves much more in water than does CH_4. Therefore, the partial permanence of H_2S in the liquid phase will imply a smaller reduction of the influent COD, when compared to the treatment of wastewaters not containing sulfate (see Section 2.3.7). According to the Buswell equation, the amount of CO_2 in the biogas can also be much smaller than expected, due to the high solubility of this gas in water.

(b)   Estimation of methane production considering the degraded COD

Another method of evaluating the production of methane is from the estimation of the COD degradation in the reactor, according to the following equation:

$$CH_4 + 2O_2 \Rightarrow CO_2 + 2H_2O \tag{2.14}$$
$$(16\ g) + (64\ g) \Rightarrow (44\ g) + (36\ g)$$

It can be concluded that one mole of methane requires two moles of oxygen for its complete oxidation to carbon dioxide and water. Therefore, every 16 grams of CH_4 produced and lost to the atmosphere corresponds to the removal of 64 grams of COD from the waste. Under normal temperature and pressure conditions, this corresponds to 350 mL of CH_4 for each gram of degraded COD. The general expression that determines the theoretical production of methane per gram of COD removed from the waste is as follows:

$$V_{CH_4} = \frac{COD_{CH_4}}{K(t)} \tag{2.15}$$

where:

V_{CH_4} = volume of methane produced (L)

COD_{CH_4} = load of COD removed from the reactor and converted into methane (gCOD)

$K(t)$ = correction factor for the operational temperature of the reactor (gCOD/L)

$$K(t) = \frac{P \times K}{R \times (273 + T)} \tag{2.16}$$

where:

P = atmospheric pressure (1 atm)

K = COD corresponding to one mole of CH_4 (64 gCOD/mole)

R = gas constant (0.08206 atm·L/mole·°K)

T = operational temperature of the reactor (°C)

Considering that the production of methane can be easily determined in an anaerobic reactor, this is a fast, direct measurement of the conversion degree of the waste and of the efficiency of the treatment system.

Example 2.1

Consider the treatment of a wastewater with the following characteristics:

- temperature: 26 °C
- flow: 500 m^3/d
- composition of the wastewater:
  sucrose ($C_{12}H_{22}O_{11}$) : C = 380 mg/L, Q = 250 m^3/d
  formic acid (CH_2O_2) : C = 430 mg/L, Q = 100 m^3/d
  acetic acid ($C_2H_4O_2$) : C = 980 mg/L, Q = 150 m^3/d

Determine:

(a) The final concentration of the wastewater in terms of COD:

By balancing the oxidation reactions of each of the compounds of the wastewater:

- concentration of COD in the sucrose
  $C_{12}H_{22}O_{11} + 12O_2 \Rightarrow 12CO_2 + 11\,H_2O$
  342 g..........384 gCOD
  380 mg/L.........x gCOD $\Rightarrow$ x = 427 mgCOD/L

- COD load due to the sucrose
  250 m^3/d $\times$ 0.427 kgCOD/m^3 = 106.8 kgCOD/d

** Example 2.1 (Continued) **

- concentration of COD in the formic acid
  $CH_2O_2 + 0.5O_2 \Rightarrow CO_2 + H_2O$
  46 g...........16 gCOD
  430 mg/L......x gCOD $\Rightarrow$ x = 150 mgCOD/L

- COD load due to the formic acid
  100 m^3/d $\times$ 0.150 kgCOD/m^3 = 15.0 kgCOD/d

- concentration of COD in the acetic acid
  $C_2H_4O_2 + 2O_2 \Rightarrow 2CO_2 + 2H_2O$
  60 g..............64 gCOD
  980 mg/L........x gCOD $\Rightarrow$ x = 1.045 mgCOD/L

- COD load due to the acetic acid
  150 m^3/d $\times$ 1.045 kgCOD/m^3 = 156.8 kgCOD/d

- final concentration of the waste in terms of COD
  Final concentration = Total load/total flow = (106.8 + 15.0 +
  156.8 kgCOD/d)/500 m^3/d
  Final concentration = (278.6 kgCOD/d)/(500 m^3/d) = 0.557 kgCOD/m^3
  (557 mgCOD/L)

(b) The maximum theoretical methane production, assuming the following
yield coefficients for acidogenic and methanogenic organisms: $Y_{acid} = 0.15$
and $Y_{methan} = 0.03$ gCOD$_{cel}$/gCOD$_{remov}$.
 The maximum theoretical production occurs when the removal efficiency of
COD is 100%, and there is no sulphate reduction in the system.

- COD load removed in the treatment system:
  278.6 kgCOD/d (100% efficiency)

- COD load converted into acidogenic biomass:
  $COD_{acid} = Y_{acid} \times 278.6 = 0.15 \times 278.6 = 41.2$ kgCOD/d

- COD load converted into methanogenic biomass:
  $COD_{methan} = Y_{methan} \times (278.6 - 41.2) = 0.03 \times 237.4 = 7.1$ kgCOD/d

- COD load converted into methane:
  COD_{CH_4} = total load $-$ load converted into biomass = 278.6 $-$ 41.2 $-$
  7.1 = 230.3 kgCOD/d

- Estimated production of methane:
  The value of K(t) is determined from Equation 2.16.
  K(t) = (P $\cdot$ K)/[R $\cdot$ (273 + t)] = (1 atm $\times$ 64 gCOD/mole)/[0.0821 atm$\cdot$
  L/mole $\cdot$ K $\times$ (273 + 26 °C)]
  K(t) = 2.61 gCOD/L

Example 2.1 (*Continued*)

The theoretical production of methane is determined from Equation 2.15.
$V_{CH_4} = COD_{CH_4}/K(t) = (230.3 \text{ kgCOD/d})/(2.61 \text{ kgCOD/m}^3)$
$V_{CH_4} = \mathbf{88.2 \ m^3/d}$

Note: The theoretical production of methane can also be calculated from Equation 2.11. In this case, the theoretical production should be calculated separately for each of the three compounds present in the wastewater, in terms of their concentrations and individual loads removed (not in terms of COD). After that, the following should be done:

- convert the methane load produced into the equivalent COD load (Equation 2.14)
- deduct the COD load converted into acidogenic and methanogenic biomass (as above)
- estimate the volumetric production of methane (Equations 2.15 and 2.16).

2.3.7 Sulfate reduction and methane production

As analysed in Section 2.2, the presence of sulfate in wastewater causes a change in the metabolic pathways in the anaerobic digester (Figure 2.2), in view of a competition for substrate established between the sulfate-reducing bacteria and the fermentative, acetogenic and methanogenic microorganisms. Hence, two final products are formed: methane (by methanogenesis) and sulfide (by sulfate reduction). The magnitude of this competition is related to several aspects, particularly the pH and the COD/SO_4^{2-} ratio in the wastewater. The production of sulfides may cause serious problems during the treatment of these wastewaters (adapted from Lettinga, 1995; Visser, 1995):

- The reduced SO_4^{2-} results in the formation of H_2S, an inhibiting compound for the methanogenic microorganisms that can reduce their activity and the capacity of the anaerobic reactor. In practice, the methanogenic microorganisms become more inhibited only when the COD/SO_4^{2-} ratio is less than 7, but are strongly dependent on the pH. For high COD/SO_4^{2-} ratios (>10), a large portion of the H_2S produced will be removed from the liquid phase, in view of a higher production of biogas, thus reducing its inhibiting effect on the liquid phase.
- Part of the hydrogen sulfide produced passes to the gaseous phase (biogas), which may cause corrosion and bad odour problems. If the biogas is intended to be used, an additional cost should be estimated for its purification.
- The presence of sulfide causes a high demand for oxygen in the effluent, as well as bad odour problems. A post-treatment phase for sulfide removal may be necessary.
- For the same amount of organic material present in the waste, the sulfate reduction decreases the amount of methane produced. A reduction of 1.5 g

of SO_4^{2-} corresponds to the use of 1.0 g of COD, which means a smaller availability for conversion into CH_4 (see Equation 2.17).

The COD used for reduction of the sulfate present in the wastewater can be estimated by the following equation:

$$S^{2-} + 2O_2 \Leftrightarrow SO_4^{2-} \qquad (2.17)$$
$$(32\ g) + (64\ g) \Rightarrow (96\ g)$$

It is noted that 1 mole of SO_4^{2-} requires two moles of oxygen for its reduction to sulfide. Therefore, every 96 g of SO_4^{2-} present in the waste consume 64 g of COD (1.5 SO_4^{2-}:1.0 COD ratio).

2.4 ENVIRONMENTAL REQUIREMENTS

2.4.1 Preliminaries

A natural habitat does not imply an environment unaffected by human activities, but an environment where the species that make up the microbial population are those selected by interaction with the environment and among themselves. Nutritional and physical conditions enable the selection of the organisms better adapted to the environment, which may vary quickly and frequently due to changes in the supply of nutrients or in the physical conditions.

Both physical and chemical characteristics of the environment influence microbial growth. Physical factors usually act as selective agents, while chemical factors can or cannot be selective. Some elements, such as carbon and nitrogen, which are usually required in relatively large amounts, can be very important in the selection of the prevailing species. Micronutrients, which are required in very small amounts, generally have little or no selective influence (Speece, 1986).

Anaerobic digestion is particularly susceptible to the strict control of the environmental conditions, as the process requires an interaction between fermentative and methanogenic organisms. A successful process depends on an accurate balance of the ecological system. Special attention should be given to the methanogenic microorganisms, as they are considered highly vulnerable to changes in the environmental conditions. The main environmental requirements of anaerobic digestion are commented below (Speece, 1983).

2.4.2 Nutrients

The nutritional needs of the microbial populations involved in biological wastewater treatment processes are usually established from the chemical composition of the microbial cells. As the precise composition is rarely known, the nutrient requirements are determined based on the empirical composition of the microbial cells. Such consideration is based on the fact that almost all living cells are formed by similar types of compounds, and that such cells present similar chemical composition, requiring therefore the same elements in the same relative proportions.

Table 2.2. Chemical composition of the methanogenic microorganisms

Macronutrients		Micronutrients	
Element	Concentration (g/kg TSS)	Element	Concentration (mg/kg TSS)
Nitrogen	65	Iron	1,800
Phosphorus	15	Nickel	100
Potassium	10	Cobalt	75
Sulfur	10	Molybdenum	60
Calcium	4	Zinc	60
Magnesium	3	Manganese	20
		Copper	10

Source: Lettinga *et al.* (1996)

The chemical composition of the methanogenic microorganisms is presented in Table 2.2.

According to Lettinga *et al.* (1996), the minimum nutrient requirements can be calculated by the following expression:

$$Nr = S_0 \cdot Y \cdot N_{bac} \cdot \frac{TSS}{VSS} \tag{2.18}$$

where:

Nr = nutrient requirement (g/L)
S_0 = concentration of influent COD (g/L)
Y = yield coefficient (gVSS/gCOD)
N_{bac} = concentration of nutrient in the bacterial cell (g/gVSS)
TSS/VSS = total solids/volatile solids ratio for the bacterial cell (usually 1.14)

For biological treatment processes to be successful, the inorganic nutrients necessary for the growth of microorganisms should be supplied in sufficient amounts. If the ideal concentration of nutrients is not supplied, there should be some form of compensation, either by applying smaller loads to the treatment system, or by allowing a reduced efficiency of the system. The presence or absence of micronutrients in the wastewater is generally evaluated by a laboratory survey. Sometimes, the combined treatment of several types of wastewater can compensate for the lack of micronutrients in some wastes.

Domestic sewage generally presents all appropriate types of nutrients in suitable concentrations, thus providing an ideal environment for the growth of microorganisms, with no limitations for the anaerobic digestion process. A possible exception is the availability of sufficient iron in sludge generated in domestic sewage treatment, which may limit the methanogenic activity. On the other hand, industrial effluents are more specific in composition and may require a nutrient supplementation for an ideal degradation.

The following nutrients, in decreasing order of importance, are necessary for the nutritional stimulation of methanogenic microorganisms: nitrogen, sulfur, phosphorus, iron, cobalt, nickel, molybdenum, selenium, riboflavin and vitamin B12.

(a) Nitrogen

Generally, nitrogen is the inorganic nutrient required in larger concentrations for the growth of microorganisms. Under anaerobic conditions, nitrogen in the forms of nitrite and nitrate is not available for bacterial growth, as it is reduced to nitrogen gas and released to the atmosphere. Ammonia and the fraction of organic nitrogen released during degradation are the main sources of nitrogen used by microorganisms.

As bacteria grow much more in wastes containing large amounts of carbohydrates than they do in wastes containing proteins and volatile acids, the nitrogen needs for the first type of waste may be about six times larger than those for the volatile acid-based wastes due to an increased synthesis of the fermentative bacteria.

Nitrogen requirements are based on the empirical chemical composition of the microbial cell, according to Table 2.2. Although an estimation of the nutrient requirements based on the sewage concentration is not the most suitable one, most of the guidelines contained in the specialised literature refer to a COD-based supplementation of nutrients. According to Lettinga *et al.* (1996), assuming that the nutrients present in sewage are in a form available to the bacteria, the following relations can be used:

- *Biomass with low yield coefficient (Y ~ 0.05 gVSS/gCOD)*
  e.g. degradation of volatile fatty acids
  COD:N:P = 1000:5:1
  C:N:P = 330:5:1

- *Biomass with high yield coefficient (Y ~ 0.15 gVSS/gCOD)*
  e.g. degradation of carbohydrates
  COD:N:P = 350:5:1
  C:N:P = 130:5:1

(b) Phosphorus

Microbial incorporation of phosphorus in anaerobic digestion has been reported as being approximately 1/5 to 1/7 of that established for nitrogen. Most of the microorganisms are capable of using inorganic orthophosphate, which can be incorporated by the growing cells by means of the mediation of enzymes named phosphatases.

(c) Sulfur

Most of the methanogenic microorganisms use sulfide as a source of sulfur, although some of them can use cysteine. If inorganic sulfate is present, it is reduced to sulfide, which then reacts with the serine amino acid to form sulfur containing the cysteine amino acid. Sulfur is necessary for the synthesis of proteins.

In general, the concentration of sulfate in domestic sewage is sufficient to provide the sulfur necessary for the bacterial growth, which is required in relatively small amounts. This is due to the low sulfur content in the microbial cells. Other reasons to disregard the need for sulfides in anaerobic digestion include: (i) presence of H_2S in the biogas; (ii) microbial synthesis of sulfide and (iii) precipitation of sulfides by metals.

Sulfur and phosphorus seem to be required in the same amount. It should be emphasised that sulfur requirements for methanogenic microorganisms are part of a complex process. On one hand, the presence of sulfates can limit the methanogenesis, because the sulfate-reducing bacteria compete for substrates such as hydrogen and acetate. On the other hand, the methanogenic organisms depend on the production of sulfides for their growth. This illustrates the relatively narrow ecological environment occupied by the methanogenic organisms, where some inorganic compounds pass from ideal to toxic concentrations within a small range.

Example 2.2

Calculate the nitrogen and phosphorus requirements of an anaerobic treatment system with the following characteristics:

- type of substrate: carbohydrate
- concentration of the influent substrate: $S_0 = 0.350$ gCOD/L
- yield coefficient: $Y = 0.15$ gVSS/gCOD
- TSS/VSS ratio of the bacterial cell: 1.14
- concentration of nutrients in the bacterial cell: 0.065 gN/gTSS; 0.015 gP/gTSS (Table 2.2)

Solution:

- Calculation of the nitrogen requirement

  Using Equation 2.18:
  Nr = 0.350 gCOD/L $\times$ 0.15 gVSS/gCOD $\times$ 0.065 gN/gTSS
       $\times$ 1.14 gTSS/gVSS
  Nr = 0.0039 gN/L (3.9 mgN/L)

- Calculation of the phosphorus requirement

  Using Equation 2.18:
  Nr = 0.350 gCOD/L $\times$ 0.15 gVSS/gCOD $\times$ 0.015 gP/gTSS
       $\times$ 1.14 gTSS/gVSS
  Nr = 0.0009 gP/L (0.9 mgP/L)

- Determination of the COD:N:P ratio

  0.350 gCOD/L:0.0039 gN/L:0.0009 gP/L
  350:3.9:0.9 or (350:4:1)

(d)   Micronutrients

Besides nitrogen, phosphorus and sulfur, which, together with carbon and oxygen, constitute the macromolecules of the microbial cells, a large number of other elements are necessary for the anaerobic digestion process. These elements are named micronutrients and comprise the micromolecules of the cells. They represent about 4% of the dry weight of the cells.

It is difficult to determine in practice the exact demand of these micronutrients, once the presence and need for sulfides by the methanogenic organisms lead to the precipitation of these elements from the solution, making the concentration of metals in equilibrium very low. To solve this situation, a pulse application of acidified influent can be performed to disturb the chemical equilibrium and make the metals momentarily available for the methanogenic microorganisms.

Iron, cobalt, nickel and molybdenum are the main micronutrients required by the microorganisms that form methane from acetate.

2.4.3 Temperature

Among the physical factors that affect microbial growth, temperature is one of the most important in the selection of species. Microorganisms are not capable of controlling their internal temperature and, consequently, the temperature inside the cell is determined by the external ambient temperature.

Three temperature ranges can be associated with microbial growth in most of the biological processes (Batstone *et al.*, 2002):

- *psycrophilic range*: between 4 and approximately 15 °C
- *mesophilic range*: between 20 and approximately 40 °C
- *thermophilic range*: between 45 and 70 °C, and above

In each of these three ranges, where microbial growth is possible, three temperature values are usually used to characterise the growth of the microorganism species (see Figure 2.6):

- *minimum* temperature, below which growth is not possible
- *optimum* temperature, in which growth is maximum
- *maximum* temperature, above which growth is also not possible

The *maximum* and *minimum* temperatures define the limits of the range in which growth is possible, and the *optimum* temperature is that in which growth rate is maximum. The microbial growth rate at temperatures close to the minimum is typically low, but it increases exponentially as the temperature rises, reaching its maximum close to the ideal temperature. From the ideal growth rate, the increase of a few degrees causes an abrupt drop in the growth rate, to zero value.

The microbial formation of methane may occur in a wide temperature range (0 to 97 °C). Two ideal temperature levels have been associated with the anaerobic digestion, one in the mesophilic range (30 to 35 °C), and another in the thermophilic

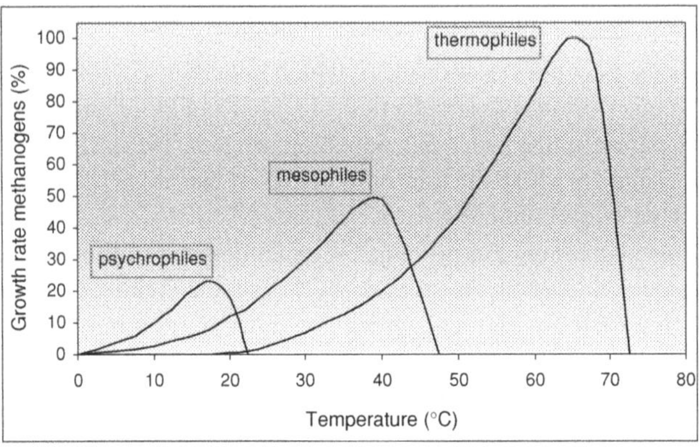

Figure 2.6.   Influence of the temperature on the biomass growth rate. Relative growth rate of psychrophilic, mesophilic and thermophilic methanogens (source: adapted from van Lier *et al.*, 1997)

range (50 to 55 °C). Most of the anaerobic digesters have been designed in the mesophilic range, although their operation is also possible in the thermophilic range. However, the operational experience of anaerobic digesters in this range has not been very satisfactory, and many questions are still pending, such as whether the resulting benefits overcome the disadvantages, including the necessary additional energy to heat up the digesters, the poor quality of the supernatant and the instability of the process.

The external effects of temperature on bacterial cells are also important. For example, the degree of dissociation of several compounds depends strongly on the temperature, such as the specific case of ammonia. The thermodynamics of several reactions is also affected by temperature, such as the dependence of the hydrogen pressure in anaerobic digesters where fermentation occurs in an appropriate manner.

The importance of the quantitative data on the effects of the temperature on the microbial population is that a considerable reduction may be achieved in the reactor volume, if it is operated close to the ideal temperature, once the maximum specific growth rate of the microbial population rises as the temperature increases. Although high temperatures are desired, maintaining a uniform temperature in the reactor may be more important, once the anaerobic process is considered very sensitive to abrupt temperature changes, which may cause an unbalance between the two largest microbial populations and the consequent failure of the process (the usual limit is about 2 °C per day).

The methane-forming microorganisms prevailing in anaerobic digesters operated in the mesophilic temperature range belong to the genera *Methanobacterium*, *Methanobrevibacter* and *Methanospirillum*, which are hydrogen-using organisms, and to the genera *Methanosarcina* and *Methanosaeta*, which are organisms that use acetate to form methane.

The temperature affects the biological processes in two ways: (i) influencing the enzymatic reaction rates; and (ii) influencing the substrate diffusion rates. Although diffusion is an important factor, particularly in full-scale reactors, only the effects of temperature related to the reaction rates are discussed in this item.

The data found in the specialised bibliography indicate that K_s and Y decrease as the temperature increases, while the K_d coefficient of low-growth-rate cultures is little affected by temperature (Grady and Lim, 1980).

The Arrhenius equation is frequently used to quantify the effects of temperature on biochemical reactions:

$$K = K_o \cdot e^{\left(\frac{-E}{R \cdot T_{abs}}\right)} \tag{2.19}$$

where:

K = reaction rate
K_o = constant
E = activation energy (cal/mole)
R = gas constant (1.98 cal/mole $\cdot$ K)
T_{abs} = absolute temperature (K)

According to the experimental data available, μ_{max} increases as the temperature rises, until a maximum growth value is reached. From this maximum value, μ_{max} decreases quickly. This decrease results from two competitive processes: (i) bacterial synthesis; and (ii) bacterial decay, each represented by the Arrhenius equation, so that the net growth rate can be expressed as follows:

$$K_{net} = K_1 \cdot e^{\left(\frac{-E_1}{R \cdot T_{abs}}\right)} - K_2 \cdot e^{\left(\frac{-E_2}{R \cdot T_{abs}}\right)} \tag{2.20}$$

where:

K_{net} = net growth rate
K_1 = bacterial synthesis rate
K_2 = bacterial decay rate

As the temperature increases, the inactivation and denaturation of enzymes and proteins become very important, until the net growth rate reaches a maximum. Above the ideal temperature, the growth rate falls suddenly, when the decay begins to prevail over synthesis.

According to Henze and Harremoës (1983), the maximum bacterial growth rate decreases 11% per °C, for anaerobic digesters operated at temperatures below 30 °C, as shown by the following expression (van Haandel and Lettinga, 1994):

$$K(t) = K_{30} \times 1.11^{(T-30)} \tag{2.21}$$

where:

$K(t)$ = growth rate for the temperature (T)
K_{30} = growth rate for t = 30 °C
T = temperature (°C)

2.4.4 pH, alkalinity and volatile acids

These three environmental factors are closely related to each other, being equally important to the control and suitable operation of anaerobic processes. The pH affects the process in two main ways (Lettinga *et al.*, 1996):

- *directly*: affecting, for example, the enzymes' activity by changing their proteic structure, which may occur drastically as a result of changes in the pH
- *indirectly*: affecting the toxicity of a number of compounds (see Section 2.5.5)

The methane-producing microorganisms have optimum growth in the pH range between 6.6 and 7.4, although stability may be achieved in the formation of methane in a wider pH range, between 6.0 and 8.0. pH values below 6.0 and above 8.3 should be avoided, as they can inhibit the methane-forming microorganisms. The optimum pH depends on the type of microorganisms involved in the digestion process, as well as on the type of substrate. Table 2.3 presents values of optimum pH ranges for the degradation of different substrates.

Regarding the stability of the process, the fact that the acid-producing bacteria are much less sensitive to pH than the methanogenic microorganisms is particularly important, as the acidogenic bacteria can still be very active, even for pH values as low as 4.5. In practice, this means that the production of acids in a reactor can continue freely, although the methane production has been practically interrupted due to the low pH values. As a result, the reactor contents will become "sour".

The acid-producing bacteria have an optimum growth rate in the pH range between 5.0 and 6.0, with a higher tolerance to lower pH values. Therefore, pH control aims mainly at eliminating the risk of inhibition of the methanogenic microorganisms by the low pH values, thus avoiding the failure of the process.

The operation of an anaerobic reactor with the pH constantly below 6.5 or above 8.0 can cause a significant decrease in the methane production rate. In addition, sudden pH changes (pH shocks) can adversely affect the process, and recovery will depend on a series of factors, related to the type of damage caused to the microorganisms (either permanent or temporary). According to Lettinga *et al.*

Table 2.3. Optimum pH ranges for the degradation of different substrates

Substrate	Optimum pH
Formiate	6.8 to 7.3
Acetate	6.5 to 7.1
Propionate	7.2 to 7.5

Source: Lettinga *et al.* (1996)

(1996), the recovery will be quicker if:

<table>
<tr><td colspan="2">Acid pH shock</td><td colspan="2">Alkaline pH shock</td></tr>
<tr><td>1.</td><td>The pH drop was not significant.</td><td>1.</td><td>The pH rise was not significant.</td></tr>
<tr><td>2.</td><td>The pH shock had a short duration.</td><td>2.</td><td>The pH shock had a short duration.</td></tr>
<tr><td>3.</td><td>The VFA concentration during the pH shock remained low.</td><td></td><td></td></tr>
</table>

(a)   Alkalinity and buffer capacity

The buffer capacity can be understood as the capacity of a solution to avoid changes in the pH. A buffer solution consists of a mixture of a weak acid and its corresponding salt, thus enabling the grouping of the ions H^+ and OH^- and avoiding both the increase and the decrease of the pH. The following generic equations are applied:

$$HA + H_2O \Leftrightarrow H_3O^+ + A^- \tag{2.22}$$

$$K_A = \frac{[H_3O^-].[A^-]}{[HA]} \tag{2.23}$$

$$pH = pK_A + \log \frac{[A^-]}{[HA]} \tag{2.24}$$

The buffer capacity reaches its maximum when $pH = pK_A$, that is, when $[A^-] = [HA]$.

The two main factors that affect the pH in anaerobic processes are carbonic acid and volatile acids. In the pH range between 6.0 and 7.5, the buffer capacity of the anaerobic system depends almost completely on the carbon dioxide/alkalinity system, which, in equilibrium with the dissociation of the carbonic acid, tends to regulate the concentration of the hydrogen ion, as explained below.

The amount of carbonic acid in solution is directly related to the amount of CO_2 in the gaseous phase, once a balance is established between the amounts of CO_2 in the liquid phase and in the gaseous phase. The portion of CO_2 dissolved in the liquid phase can be established by Henry's law:

$$[CO_2] = K_H \cdot P_{CO_2} \tag{2.25}$$

where:
   $[CO_2]$ = saturation concentration of CO_2 in water (mole)
      K_H = constant of Henry's law related to the CO_2-water balance (mole/atm·L)
   P_{CO_2} = CO_2 partial pressure (atm)

The relation between alkalinity and pH is then given by the following expression (Foresti, 1994; Lettinga *et al.*, 1996):

$$pH = pK_1 + \log \frac{[HCO_3^-]}{[H_2CO_3^*]} \tag{2.26}$$

where:

$pK_1 = \log (1/K_1)$

$K_1 =$ constant of apparent ionisation (4.45×10^{-7}, at 25 °C), that is related to all the CO_2 dissolved in the liquid

$$[H_2CO_3^*] = [CO_2] + [H_2CO_3] \cong [\sim CO_2(líq)] \tag{2.27}$$

Hence, the portion of $H_2CO_3^*$ can be obtained by calculating the partial carbon dioxide gas pressure, according to Equation 2.25.

(b)   Interaction between alkalinity and volatile acids

The interaction between alkalinity and volatile acids during anaerobic digestion is based on whether the alkalinity of the system is able to neutralise the acids formed in the process and buffer the pH in case of accumulation of volatile acids. Both the alkalinity and the volatile acids derive primarily from the decomposition of organic compounds during digestion, as follows:

- Conversion of intermediate volatile fatty acids. The digestion of sodium acetate, for example, can lead to the formation of sodium bicarbonate

$$CH_3COONa + H_2O \Rightarrow CH_4 + CO_2 + NaOH \Rightarrow CH_4 + NaHCO_3 \tag{2.28}$$

- Conversion of proteins and amino acids, with formation of ammonia (NH_4^-). The combination between ammonia and carbonic acid in solution leads to the formation of ammonia bicarbonate

$$NH_3 + H_2O + CO_2 \Rightarrow NH_4^+ + HCO_3^- \tag{2.29}$$

Digestion of other organic compounds that do not lead to a cation as final product does not produce alkalinity. This occurs, for example, in the degradation of carbohydrates and alcohols. This is particularly important due to the high microbial synthesis during the degradation of carbohydrates, which could result in the depression of alkalinity, should the present ammonia bicarbonate be used as source of nitrogen for biological synthesis.

(c)   Alkalinity of the volatile acids

As a result of the reaction of the alkalinity with the volatile fatty acids produced in the system, the bicarbonate alkalinity is converted into alkalinity of volatile acids, because volatile acids are stronger than bicarbonates. However, the alkalinity buffering capacity of the volatile acids is situated in the pH range between 3.75 and 5.75, being, therefore, of little importance in anaerobic digestion. Consequently, a supplementation of the bicarbonate alkalinity lost in the reaction with the volatile acids should be provided.

In practice, for calculation of the bicarbonate alkalinity, the portion corresponding to the alkalinity of the volatile acids should be discounted from the total alkalinity, as follows (Foresti, 1994):

$$BA = TA - 0.85 \times 0.83 \times VFA = TA - 0.71 \times VFA \qquad (2.30)$$

where:

BA = bicarbonate alkalinity (as mgCaCO$_3$/L)
TA = total alkalinity (as mgCaCO$_3$/L)
VFA = concentration of volatile fatty acids (as mg acetic acid/L)
0.85 = correction factor that considers 85% of ionisation of the acids to the titration end point
0.83 = conversion factor from acetic acid into alkalinity

(d) Monitoring of alkalinity

In the monitoring of anaerobic reactors, the systematic verification of the alkalinity becomes more important than the evaluation of the pH. This is due to the logarithmic scale of pH, meaning that small pH decreases imply the consumption of a large amount of alkalinity, thus reducing the buffering capacity of the medium.

To determine separately the portions of bicarbonate alkalinity and of alkalinity of the volatile acids, the titration of the sample can be performed in two stages, according to the methodology proposed by Ripley *et al.* (1986):

- *titration up to pH 5.75*: the first stage of titration provides the *partial alkalinity* (PA), practically equivalent to the bicarbonate alkalinity
- *titration up to pH 4.3*: the second stage of titration provides the *intermediate alkalinity* (IA), practically equivalent to the alkalinity of the volatile acids

An important aspect of determining the alkalinity in two stages refers to the significance of the IA/PA ratio. According to Ripley *et al.* (1986), IA/PA values higher than 0.3 indicate the occurrence of disturbances in the anaerobic digestion process. The stability of the process is possible for IA/PA values different from 0.3, and the verification of each particular case is recommended (Foresti, 1994).

(e) Alkalinity necessary for the process

From the operational point of view, if the alkalinity is generated from the influent sewage, the maintenance of high levels of alkalinity in the system is desirable because high concentrations of volatile acids could be buffered without causing a substantial drop in pH. However, if an alkalinity supplementation is necessary, then the selection of chemical compounds shall be evaluated in terms of applicability and economy. The minimum acceptable alkalinity requirement depends on the concentration of the sewage, a decisive factor to determine the potential generation of acids in the system.

According to van Haandel and Lettinga (1994), the most important issue related to the pH value and stability is whether the alkalinity of the medium (influent alkalinity+generated alkalinity) is sufficient to keep itself at levels considered safe. The authors present a complete methodology, relating the determination of the pH and alkalinity in anaerobic digesters.

(f) Chemical products for alkalinity supplementation

Several chemical products can be used to control the pH of anaerobic processes, including hydrated lime ($Ca(OH)_2$), quicklime (CaO), sodium carbonate (Na_2CO_3), sodium bicarbonate ($NaHCO_3$), sodium hydroxide ($NaOH$) and ammonia bicarbonate (NH_4HCO_3). These chemical products can be separated into two groups:

- those that provide bicarbonate alkalinity directly ($NaOH$, $NaHCO_3$, NH_4HCO_3)
- those that react with carbon dioxide to form bicarbonate alkalinity (CaO, $Ca(OH)_2$, NH_3)

Lime is usually the cheapest source of alkalinity but, as it is a very insoluble product, it can cause serious operational problems. Carbon dioxide reacts with lime to form calcium bicarbonate, which can cause vacuum in closed digesters. If the carbon dioxide present is insufficient to react entirely with lime, the final pH may be very high, which can be as harmful as a very low pH. The formation of undesirable precipitates and fouling can cause serious operational problems.

Sodium bicarbonate is easy to handle, is very soluble and, unlike lime, it neither requires carbon dioxide nor increases pH substantially when excessively dosed. However, the cost of the product is very high.

The use of ammonia as a source of alkalinity depends substantially on the local conditions. For example, the use of anhydrous ammonia, in spite of it being cheap, may be prohibitive because the effluent will contain an excessive amount of ammonia. Besides that, care should be taken to prevent biomass toxicity by ammonia.

2.4.5 Toxic materials and their control

The appropriate degradation of organic sewage by any biological process depends on the maintenance of a favourable environment for microorganisms, including either the control or the elimination of toxic materials. Since any compound present in sufficiently high concentrations can be toxic, the toxicity should be discussed in terms of toxic levels, instead of toxic materials. In this aspect, according to Speece *et al.* (1986), the following considerations are pertinent:

- What are the required concentrations that cause toxicity?
- Is the toxic effect reversible or bactericide?
- What is the acclimatisation potential of the microorganisms?

Toxicity has been considered one of the main reasons for a non-generalised use of anaerobic digestion, once there is a widespread understanding that anaerobic processes are not capable of tolerating toxicity. It is true that methanogenic microorganisms can be more easily inhibited by toxins, due to the relatively small fraction of substrate converted into cells and to the long generation period of these microorganisms. However, microorganisms usually have a certain capacity of adaptation to the inhibiting concentrations of most of the compounds, provided that the toxicity impact minimised by some design measures, such as long solids retention time and minimised residence time of toxins in the system. The following control methods for toxic materials were suggested by McCarty (1964):

- removal of the toxic materials present in the sewage
- dilution below the toxic limit
- formation of insoluble complexes or precipitation
- antagonism of toxicity by means of the use of another compound

Several organic and inorganic compounds can be toxic or inhibitors to the anaerobic process, although the general effect resulting from the addition of most of them may vary from stimulating to toxic. Microbial activity is usually stimulated at low concentrations, but it also depends on the type of compound present. As the concentration is increased, inhibition may become high, and the rate of microbial activity may fall to zero.

(a)   Toxicity by salts

Toxicity by salts is usually associated with the cation, and not with the anion of the salt. Cation toxicity assessments carried out by Kugelman and McCarty (1965) indicated the following increasing order of inhibition, based on the molar concentration: Na^+ (0.32 M), NH_4^+ (0.25 M), K^+ (0.15 M), Ca^{2+} (0.11 M) and Mg^{2+} (0.08 M). However, more recent studies have shown that the inhibiting concentrations can be higher, provided that the biomass undergoes an adaptation stage (Lettinga *et al.*, 1996).

If some cation is found at an inhibiting concentration in the influent sewage, inhibition can be reduced if an antagonistic ion is either present or added to the system. Sodium and potassium are the best antagonists for that purpose, provided that they are used in stimulating concentrations, as indicated in Table 2.4. Antagonistic elements are usually added by means of chloride salts.

Table 2.4. Stimulating and inhibiting concentrations of some cations

Cation	Concentration (mg/L)		
	Stimulating	Moderately inhibiting	Strongly inhibiting
Calcium	100 to 200	2,500 to 4,500	8,000
Magnesium	75 to 150	1,000 to 1,500	3,000
Potassium	200 to 400	2,500 to 4,500	12,000
Sodium	100 to 200	3,500 to 5,500	8,000

Source: McCarty (1964)

(b) Toxicity by ammonia

Usually, the presence of ammonia bicarbonate, resulting from the digestion of sewage rich in urea- or protein-based compounds, is beneficial to the digester as a source of nitrogen and as a buffer for pH changes. However, both the ammonium ion (NH_4^+) and the free ammonia (NH_3) can become inhibitors when present in high concentrations. These two forms of ammonia are balanced, with the relative concentration of each depending on the pH of the medium, as indicated in the following equation:

$$NH_4^+ \Leftrightarrow NH_3 + H^+ \tag{2.31}$$

For high concentrations of hydrogen ion (pH equal to or lower than 7.2), the balance shifts to the left, so that inhibition becomes related to the concentration of the ammonium ion. For higher pH levels, the concentration of hydrogen ion decreases, and the balance shifts to the right. In this situation, free ammonia may become the inhibiting agent. Studies have shown that concentrations of free ammonia above 150 mg/L are toxic to the methanogenic microorganisms, while the maximum safety limit for the ammonium ion is approximately 3,000 mg/L. The concentrations of free ammonia that can have either a beneficial or an adverse effect on anaerobic processes are presented in Table 2.5.

(c) Toxicity by sulfide

Toxicity by sulfide is a potential problem in anaerobic treatment, firstly due to the biological reduction of sulfates and organic sulfur-containing compounds, and also for the anaerobic degradation of protein-rich compounds. As covered in Sections 2.3.6 (Equation 2.13) and 2.3.7, the reduced sulfate leads to the formation of H_2S, which dissociates in water, in accordance with the following equations (Jansen, 1995):

$$H_2S \Leftrightarrow H^+ + HS^- \tag{2.32}$$

$$HS^- \Leftrightarrow H^+ + S^{2-} \tag{2.33}$$

The dissociation of species is related to the temperature and to the pH of the medium, in accordance with the distribution diagram shown in Figure 2.7,

Table 2.5. Effects of free ammonia on anaerobic processes

Concentration (as N, mg/L)	Effect
50 to 200	Beneficial
200 to 1,000	No adverse effect
1,500 to 3,000	Inhibitor for pH > 7.4 to 7.6
Above 3,000	Toxic

Source: McCarty (1964)

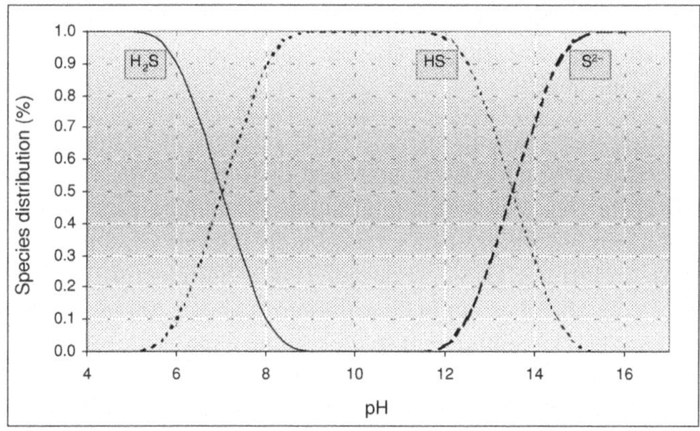

Figure 2.7.   Distribution diagram for H_2S (T = 25 °C)

developed for a temperature of 25 °C. From the analysis of the diagram, it can be concluded that:

- the un-ionised form (H_2S) is the main dissolved component for pH values lower than 7
- the ionised form (HS^-) prevails for pH values between 7 and 14
- the concentration of free sulfide (S^{2-}) is negligible in the pH range associated with sewage treatment

Inhibition by sulfide is dependent on the concentration of non-dissociated hydrogen sulfide (H_2S) in the medium, which indicates that the inhibition by sulfide is strongly dependent on pH, within the pH range usually associated with anaerobic digestion (6.5 to 8). The distribution diagram shows that, for a pH value equal to 7, around 50% of the sulfide will be present in the most toxic, non-dissociated form (H_2S) and the other 50% in the less toxic, dissociated form (HS^-). On the other hand, H_2S can still be either present in the gaseous phase (H_2S_{gas}) or dissolved in the liquid phase (H_2S_{liq}). The higher or lower presence of sulfides in the gaseous phase will strongly depend on the gas production in the system. The greater the production of CH_4 in the reactor, the larger the amount of sulfides in the gaseous form removed from the liquid phase. Consequently, the toxicity of H_2S will decrease as the concentration of influent COD increases (larger production of CH_4). It is generally assumed that, for a COD/SO_4^{2-} ratio higher than 10, toxicity problems will not occur in the anaerobic reactor.

From the practical point of view, it is important to determine the sensitivity of the biomass to sulfide. The amount of sulfides produced in the anaerobic treatment depends on the following main factors:

- COD/SO_4^{2-} ratio in the influent (a low ratio results in a high sulfide production)
- composition of the organic substrate

- pH and temperature of the medium
- result of the competition between sulfate-reducing and methanogenic microorganisms

For the design and operation of anaerobic reactors, it is important to know the maximum allowable concentration of non-dissociated H_2S. According to the literature, anaerobic reactors with a high biomass retention capacity (e.g. UASB reactors and anaerobic filters) can tolerate higher levels of sulfide, amounting approximately to 170 mg H_2S/L (Speece, 1986). Sulfides in the form of H_2S become very toxic when present in concentrations above 200 mg/L, but they can be tolerated up to this concentration if the operation of the system is continuous and if the biomass undergoes some acclimatisation. Sulfide concentrations amounting to 50 to 100 mg/L can be tolerated with little or no system acclimatisation.

If the sulfide concentration in the reactor exceeds the maximum tolerable values, special measures should be taken to ensure a good performance of the system:

- increase pH in the reactor, so that the dissociation of H_2S in the liquid phase favours the formation of HS^-. From Figure 2.7, only 10% of the sulfide will be present in non-dissociated form if the pH in the reactor is equal to 8
- dilute the influent, aiming at reducing the concentration of sulfides in the reactor
- precipitate sulfides by using iron salts
- increase COD/SO_4^{2-} ratio, to favour the release of H_2S from the liquid phase to the gaseous phase

(d)   Toxicity by metals

Toxic elements and compounds such as chromium, chromates, nickel, zinc, copper, arsenic and cyanides, among others, are classified as highly toxic inorganic toxins. In particular, the presence of low concentrations of copper, zinc and nickel in soluble state is considered highly toxic, and these salts are associated with most of the toxicity problems caused by metals in anaerobic treatment.

The concentrations of the most toxic metals that can be tolerated in anaerobic treatment are related to the concentrations of sulfide available to be combined with the metals and then form insoluble sulfide salts. Sulfides by themselves are very toxic to anaerobic treatment but, when combined with metals, they form insoluble salts that have no adverse effect.

One of the most effective procedures to control toxicity by metals is the addition of sufficient amounts of sulfide to precipitate the metals. Approximately 1.8 to 2.0 mg/L of metals is precipitated as metallic sulfides by the addition of 1.0 mg/L of sulfide (S^{2-}). This phenomenon is a good alternative for the treatment of industrial effluents containing metals. If this ratio (1 mg/L of sulfide:2 mg/L of metals) is not verified during the treatment, the addition of sodium sulfide or of a sulfate salt is recommended.

3

Biomass in anaerobic systems

3.1 PRELIMINARIES

A biological treatment process tends to be economical if it can be operated at low hydraulic detention times and at sufficiently long solids retention times to allow microorganism growth. This was for many years the greatest problem of anaerobic digestion, as the solids retention time could not be controlled independently of the hydraulic detention time. Thus, the microorganisms involved in the process, which have low growth rates, needed extremely long retention times and consequently reactors of large volumes. The development of high-rate anaerobic processes solved this problem, since these processes are capable of allowing the presence of a large amount of high-activity biomass, which can be maintained in the reactor even when operated at low hydraulic detention times. If sufficient contact can be guaranteed between the biomass and the organic compounds, high volumetric loads can then be applied to the system.

3.2 BIOMASS RETENTION IN ANAEROBIC SYSTEMS

3.2.1 Preliminaries

Microbial cells exist in a wide range of sizes, forms and growth phases, individually or aggregated in several microstructures. These conditions have a practical meaning in anaerobic digestion, as it is probable that the biomass form has a significant effect on the survival of the organisms and on the transfer of nutrients and, consequently, on the global efficiency of the anaerobic digestion process.

The formation of a certain structure of aggregated cells depends on several factors, including the size range of the cells inside the microbial population and the location of each individual cell in relation to the others and to the growth medium, for example in the gas/liquid interface. The retention of high-activity biomass in high-rate anaerobic processes depends on a series of factors and mechanisms, as discussed in the following items (adapted from Stronach *et al.*, 1986).

3.2.2 Retention by attachment

The habitats of microorganisms in aqueous systems, such as anaerobic digesters, are very diverse, and their survival and growth depend on factors such as temperature, nutrient availability and stratification. The organisms often overcome the instability of the environment where they live by attachment to a surface. The attachment capability of bacteria is impressive. Their superficial structures seem to allow some form of control of the adhesion, while their microscopic dimensions guarantee that they are hardly subjected to the shearing forces that happen naturally in the medium.

This form of immobilisation of microorganisms, through attachment, is possible on fixed surfaces, such as in anaerobic processes with a stationary bed (e.g. anaerobic filter), or on moving surfaces, such as in anaerobic processes of expanded and fluidised beds. Figure 3.1 illustrates the biofilm formation attached to a support medium.

3.2.3 Retention by flocculation

Flocculation has a practical meaning in sewage treatment, since the flocculating microstructures can be easily separated from the liquid phase by sedimentation. The phenomenon of flocculation is particularly important in two-stage processes and in upflow anaerobic sludge blanket (UASB) reactors. Bacterial growth in flocs is not necessary for an efficient substrate removal, but it is essential to guarantee an effluent with low concentrations of suspended solids.

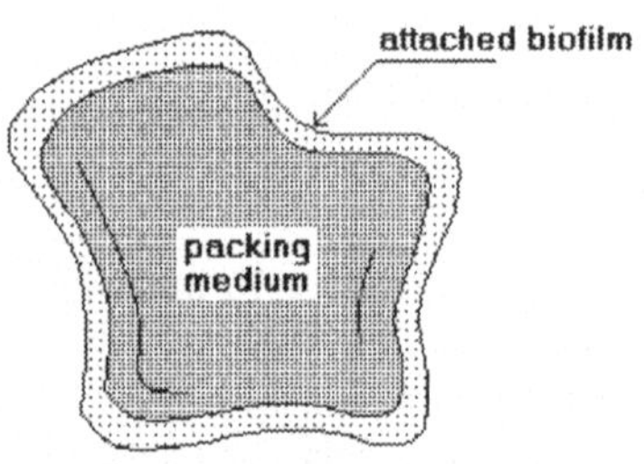

Figure 3.1.   Biomass retention by attachment

3.2.4 Retention by granulation

In terms of wastewater treatment, the phenomenon of granulation (formation of granules) seems to be restricted to UASB reactors (and its variants) and, to a lesser extent, to anaerobic filters. This is usually associated with the treatment of wastewaters rich in carbohydrates and volatile acids.

The mechanisms that control the selection and formation of granules are related to physical, chemical and biological factors, including (Lettinga *et al.*, 1980; Hulshoff Pol *et al.*, 1984; Wiegant and Lettinga, 1985):

- the characteristics of the substrate (concentration and composition)
- the gravitational compression of the sludge particles and the superficial rate of biogas liberation
- the ideal conditions for the growth of the methanogenic archaea, such as the presence of bivalent cations
- the upflow velocity of the liquid through the sludge bed

Particularly important is the upflow velocity of the liquid, which provides a constant selective pressure on the microorganisms that start adhering to each other and thereby leads to the formation of granules that present good settleability. The granules usually have a well-defined appearance and they can be several millimetres in diameter and accumulate in large amounts in the reactor. The granular configuration presents several advantages from an engineering point of view (Guiot *et al.*, 1992):

- the microorganisms are usually densely grouped
- the non-use of inert support mediums enables the maximum use of the reaction volume of the reactor
- the spherical form of the granules provides a maximum microorganism/ volume ratio
- the granules present excellent settleability

In the arrangement of biomass in granules, the different bacterial populations seem to selectively group in layers on top of each other, for example like the model proposed by Guiot *et al.* (1992) for the substrate and product diffusion (Figure 3.2).

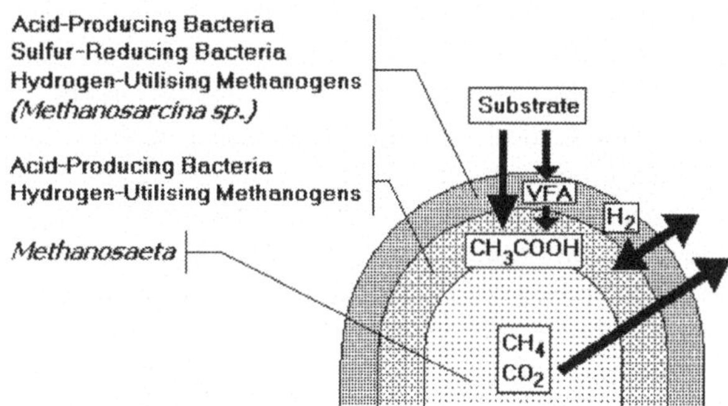

Figure 3.2.   Microorganism structure in a granule (after Guiot *et al.*, 1992)

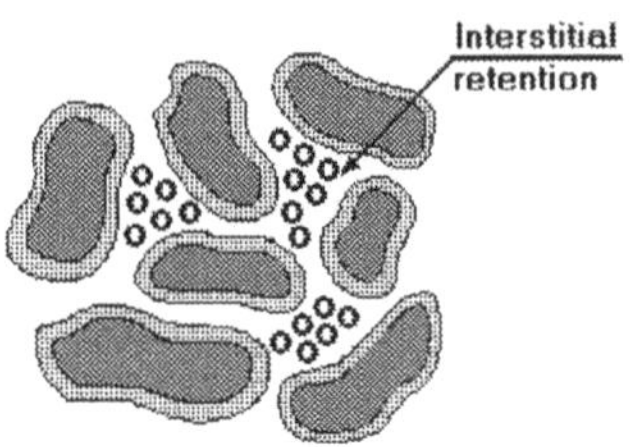

Figure 3.3.   Interstitial biomass retention

3.2.5 Interstitial retention

This type of biomass immobilisation occurs in the interstices (Figure 3.3) of stationary support mediums, as is the case of fixed bed anaerobic reactors. The surfaces of the medium serve as support for the attached bacterial growth (formation of the biofilm), while the empty spaces in the packing material are occupied by microorganisms that grow dispersely.

3.3 EVALUATION OF THE MICROBIAL MASS

The determination of the biomass in anaerobic digesters presents two main difficulties: (i) in some systems, the microorganisms are attached to small inert particles; and (ii) the biomass is usually present as a consortium of different morphologic and physiologic types.

The determination of the biomass and the microbial composition usually requires the extraction, isolation and separation of the biochemical constituents that are specific to a certain group of microorganisms. The cellular components that change quickly in nature, after the death of a cell, can be used, for example, for the estimation of the viable biomass.

Although there are several methodologies to evaluate the amount and activity of the biomass in anaerobic digesters, most of them are sophisticated and cannot be adopted as control and monitoring parameters for reactors operating in full scale, especially if considering the existing laboratory resources in many developing countries. Hence, the evaluation of the amount of biomass is usually made through the determination of the vertical solids profile, considering that the volatile solids are a measure of the biomass present in the reactor (mass of cellular material). Sludge samples collected at different levels of the reactor height are gravimetrically analysed and the results are expressed in terms of grams of volatile solids per litre (gVS/L). These concentration values of volatile solids (made for each of the sludge sampling points along the reactor height), multiplied by the volumes corresponding to each sampled zone, provide the mass of microorganisms along the reactor profile. The sum of the biomass quantities in each zone is equal to the total mass of solids in the reactor, as shown in Example 3.1.

Example 3.1

Determine the amount and the average concentration of the biomass in an anaerobic reactor. Data are:

- total reactor volume: $V = 1{,}003.5$ m^3
- volume of the digestion compartment: $V_{dc} = 752.6$ m^3
- volume of the sedimentation compartment: $V_{sc} = 250.9$ m^3
- volumes corresponding to each sampled zone, as indicated in the illustration below (V_1 to V_5)
- sludge concentration in each sampled zone, as indicated in the illustration below (C_1 to C_5)

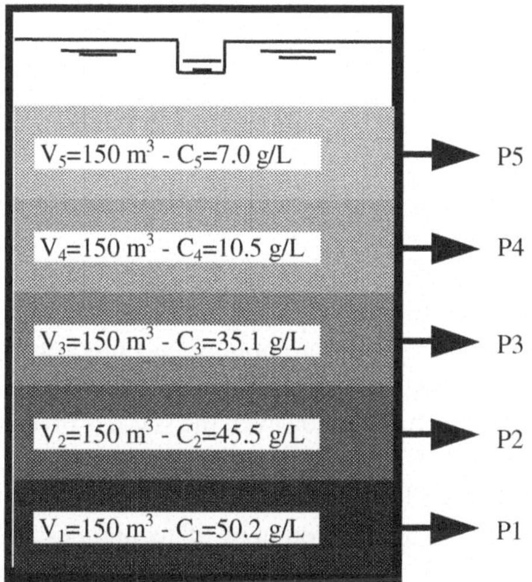

Solution:

- Calculation of the amount of biomass (M) in each zone of the reactor:

$$\text{Zone 1: } M_1 = C_1 \times V_1 = 50.2 \text{ kgVS/m}^3 \times 150 \text{ m}^3 = 7{,}530 \text{ kgVS}$$
$$\text{Zone 2: } M_2 = C_2 \times V_2 = 45.5 \text{ kgVS/m}^3 \times 150 \text{ m}^3 = 6{,}750 \text{ kgVS}$$
$$\text{Zone 3: } M_3 = C_3 \times V_3 = 35.1 \text{ kgVS/m}^3 \times 150 \text{ m}^3 = 5{,}265 \text{ kgVS}$$
$$\text{Zone 4: } M_4 = C_4 \times V_4 = 10.5 \text{ kgVS/m}^3 \times 150 \text{ m}^3 = 1{,}575 \text{ kgVS}$$
$$\text{Zone 5: } M_5 = C_5 \times V_5 = 7.0 \text{ kgVS/m}^3 \times 150 \text{ m}^3 = 1{,}050 \text{ kgVS}$$

- Calculation of the amount of biomass in the digestion compartment (M_{dc}):

$$M_{dc} = M_1 + M_2 + M_3 + M_4 + M_5 = 22{,}170 \text{ kgVS}$$

Example 3.1 (Continued)

- Calculation of the average biomass concentration in the digestion compartment (C_{dc})

$$C_{dc} = M_{dc}/V_{dc} = 22{,}170 \text{ kgVS}/750 \text{ m}^3 = 29.6 \text{ kgVS}/\text{m}^3$$
$$= 29.6 \text{ gVS/L} = 29{,}600 \text{ mgVS/L} \approx 3.0\%$$

- Calculation of the average biomass concentration in the reactor (C_r):

Assuming that the amount of biomass in the settling compartment is negligible when compared to the digestion compartment, it can be stated that $M_r = M_{dc}$

$$C_r = M_r/V = 22{,}170 \text{ kgVS}/1{,}003.5 \text{ m}^3 = 22.1 \text{ kgVS}/\text{m}^3$$
$$= 22.1 \text{ gVS/L} = 22{,}100 \text{ mgVS/L} \approx 2.2\%$$

3.4 EVALUATION OF THE MICROBIAL ACTIVITY

3.4.1 Preliminaries

In the last few years, with the development of high-rate anaerobic processes and the increased knowledge of the microbiology and biochemistry of the process, a growing use of anaerobic digestion has been observed for the treatment of a diverse number of liquid effluents. However, the success of any anaerobic process, especially the high-rate ones, depends fundamentally on the maintenance (inside the reactors) of an adapted biomass with a high microbiological activity that is resistant to shock loads. The development of techniques for the evaluation of the microbial activity in anaerobic reactors is very important, especially of the methanogenic archaea, so that the biomass can be preserved and monitored.

In this respect, several methods have been proposed to evaluate the anaerobic microbial activity, considering the assessment of the **specific methanogenic activity (SMA)**. However, the precision of several methodologies was considered doubtful or too sophisticated for reproduction in laboratories. Another problem identified refers to the difficulty, or even impossibility, in obtaining anaerobic sludge in sufficient amounts, from reactors in laboratory scale, for the development of conventional tests.

A preliminary analysis of the studies already developed in the area indicates that some methods used for the evaluation of the SMA are crude or imprecise, whilst others are too expensive or sophisticated. The simplified method developed by James *et al.* (1990), from an adaptation of the operation of the Warburg respirometer, was undoubtedly a valuable contribution, but as the authors themselves stated, greater success was dependent on the automation of the gas measurement system and on the optimisation of the monitoring system of the test as a whole.

In this regard, the work developed by Monteggia (1991), incorporating manometers with electric sensors for the continuous monitoring of the biogas production, constituted an important improvement on the SMA test.

Recently, some innovations have been presented in relation to the gas measurement system, which replaced the conventional manometers with pressure transducers. The incorporation of these devices facilitated significantly the detection of the pressure differential inside the reaction and control flasks, besides allowing the transmission of electric pulses to a computer terminal.

3.4.2 Importance of the SMA test

The evaluation of the specific methanogenic activity of anaerobic sludge has proved important in the effort to classify the biomass potential in the conversion of soluble substrate into methane and carbon dioxide. The microbial activity test can be used as a routine analysis to quantify the methanogenic activity of anaerobic sludge or, also, in a series of other applications, as listed below:

- to evaluate the behaviour of biomass under the effect of potentially inhibiting compounds
- to determine the relative toxicity of chemical compounds present in liquid effluents and solid residues
- to establish the degree of degradability of several substrates, especially of industrial wastewater
- to monitor the changes of activity of the sludge, because of a possible accumulation of inert materials after long periods of reactor operation
- to determine the maximum organic load that can be applied to a certain sludge type, providing an acceleration of the start-up stage of treatment systems
- to evaluate kinetic parameters

3.4.3 Brief description of the SMA test

In practice, the SMA test consists in the evaluation of the capacity of the methanogenic archaea to convert organic substrate into methane and carbon dioxide gas. Thus, from known amounts of biomass (gVS) and substrate (gCOD), and under established conditions, the production of methane can be evaluated during the test period. The SMA is calculated based on the maximum methane productivity rates ($mLCH_4/gVS \cdot h$ or $gCOD\text{-}CH_4/gVS \cdot d$). The conversion of $mLCH_4$ into $gCOD\text{-}CH_4$ is done according to Equations 2.15 and 2.16 (Chapter 2). For the development of the test, the following are necessary:

- anaerobic sludge, for which the SMA is to be evaluated
- organic substrate (usually sodium acetate is used)
- buffer and nutrient solution (see Table 3.1)
- reaction flasks

Table 3.1. Buffer and nutrient solution

Solution	Reagent	Concentration	Purpose
1	KH_2PO_4	1,500 mg/L	Buffer
	K_2HPO_4	1,500 mg/L	
	NH_4Cl	500 mg/L	Macronutrient
	$Na_2S \cdot 7H_2O$	50 mg/L	
2	$FeCl_3 \cdot 6H_2O$	2,000 mg/L	
	$ZnCl_2$	50 mg/L	
	$CuCl_2 \cdot 4H_2O$	30 mg/L	
	$MnCl_2 \cdot 2H_2O$	500 mg/L	Micronutrient
	$(NH_4)_6 \cdot Mo_7O_{24}4H_2O$	50 mg/L	
	$AlCl_3$	50 mg/L	
	$CoCl_3 \cdot 6H_2O$	2,000 mg/L	
	HCl (concentrated)	1 mL	

Note: At the time solutions are used, add 1 mL of solution 2 per litre of solution 1 to obtain a single solution that shall be added to the reaction flask.
Source: Monteggia (1991)

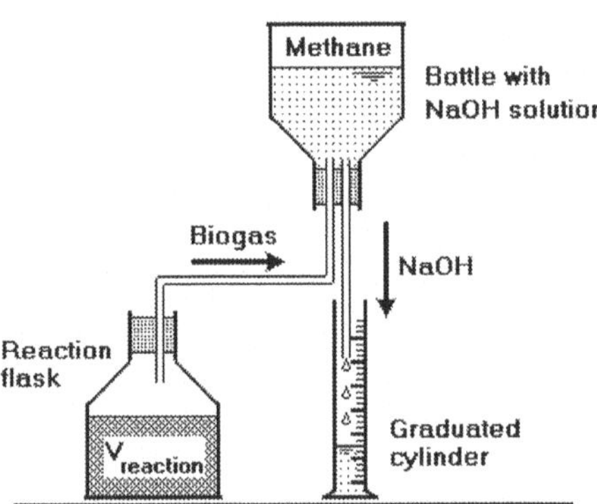

Figure 3.4.   Apparatus for biogas measurement (adapted from van Haandel and Lettinga, 1984)

- temperature controlling device (water bath, incubator, heat apparatus, acclimatised room, etc.)
- mixing device for the sludge sample
- device for measuring gas production over a certain period of time. The measurement of the production of gases can be evaluated in different ways, each with its advantages and disadvantages:
  - through water displacement (see Figure 3.4)
  - through mini-manometers (visual reading or with an electric sensor)
  - through pressure transducers etc.

Although there are different methods to follow in the development of SMA tests, the following protocol for the test was recently adopted by PROSAB (Brazilian

Research Programme on Basic Sanitation):

- determine the concentration of volatile solids present in the sludge to be analysed (gVS/L)
- place the pre-established amounts of sludge into the reaction flasks, preferably 12 to 24 hours before the beginning of the test, seeking to adapt them to the test conditions. Reaction flasks of 250 to 500 mL have usually been used at a temperature of 30 °C for the development of the test
- add to the reaction flasks certain amounts of the buffer and nutrient solution, to obtain final concentrations of the mixture (sludge+solution+substrate) of around 2.5 gVS/L. The final volume of the mixture should occupy between 70 and 90% of the volume of the reaction flask
- before adding the substrate, the oxygen present in the head space of the flask should be removed using gaseous nitrogen (pressure of 5 psi, for 5 minutes)
- add the substrate to the reaction flasks, in the concentrations desired (usually with concentrations varying from 1.0 to 2.5 gCOD/L)
- turn on the mixing device in the reaction flasks
- record the volumes of biogas produced at each time interval, during the test period (mL/hour). The determination of the methane concentration in the biogas can be made by chromatography or, alternatively, by the absorption of the carbon dioxide gas present in the biogas, through its passage in an alkaline solution (e.g. NaOH 5%)

Example 3.2

Determine the main parameters necessary for the development of a SMA test of an anaerobic sludge, considering:

- number of reaction flasks: 4
- test temperature: $T = 30$ °C
- volume of each reaction flask: 250 mL
- total volume of the mixture (sludge+solution+substrate): 200 mL (20% head space)
- concentration of the anaerobic sludge to be tested: 3% (30 gVS/L)
- sludge concentration in the mixture (sludge+solution+substrate): 2.5 gVS/L
- COD concentrations tested (gCOD/L): 1.0 (flask 1), 1.5 (flask 2), 2.0 (flask 3) and 2.5 (flask 4)

Solution:

- Determination of the sludge volume to be added to each flask, to obtain the final concentration in the mixture (sludge+solution+substrate)

Example 3.2 (Continued)

equal to 2.5 gVS/L:

$$V_{sludge} = (V_{mixture} \times C_{mixture})/C_{sludge} = (200\,mL \times 2.5\,gVS/L)/30\,gVS/L$$

$$= 16.7\,mL$$

- Determination of the mass of microorganisms in each flask:

$$M_{sludge} = V_{sludge} \times C_{sludge} = 16.7\,mL \times 0.030\,gVS/mL = 0.501\,gVS$$

- Determination of the substrate volume to be added to each flask, to obtain the final concentrations of 1.0, 1.5, 2.0 and 2.5 gCOD/L

  Considering the application of the sodium acetate solution with a concentration of 100 gCOD/L:
  - flask 1 (1.0 gCOD/L): $V_{substrate} = (C_{mixture} \times V_{mixture}) / C_{solution} = (1.0\,mgCOD/mL \times 200\,mL)/100\,mgCOD/mL = 2\,mL$
  - flask 2 (1.5 gCOD/L): $V_{substrate} = (1.5\,mgCOD/mL \times 200\,mL)/100\,mgCOD/mL = 3\,mL$
  - flask 3 (2.0 gCOD/L): $V_{substrate} = (2.0\,mgCOD/mL \times 200\,mL)/100\,mgCOD/mL = 4\,mL$
  - flask 4 (2.5 gCOD/L): $V_{substrate} = (2.5\,mgCOD/mL \times 200\,mL)/100\,mgCOD/mL = 5\,mL$

- Determination of the volume of buffer and nutrient solution:

Knowing that the total volume of the mixture was established at 200 ml, the volume of buffer and nutrient solution can be obtained by subtracting the sludge and substrate volumes already calculated from the total volume (see the following table).

Flask	Sludge concentration (gVS/L)	Volume (mL) Sludge	Volume (mL) Substrate	Volume (mL) Solution	Volume (mL) Mixture	Quantity of biomass (gVS)	Final concentration Sludge (gVS/L)	Final concentration Substrate (gCOD/L)
1	30	16.7	2	181.3	200	0.501	2.5	1.0
2	30	16.7	3	180.3	200	0.501	2.5	1.5
3	30	16.7	4	179.3	200	0.501	2.5	2.0
4	30	16.7	5	178.3	200	0.501	2.5	2.5

Once the preparatory parameters for the test have been defined, as shown in the above table, one should proceed according to the test protocol described in Section 3.4.3. The continuous monitoring of the methane production in the reaction flasks makes it possible to obtain data that correlate time with cumulative CH_4 production. The graphic representation of these data allows obtaining curves similar to those presented in Figure 3.5, one for each of the reaction flasks (1 to 4).

The determination of the specific methanogenic activity is done based on the evaluation of the slope of the line of best fit of the methane production curve (steepest reach). The slope gives the methane production rate (e.g. $mLCH_4$/hour) which, divided by the initial amount of biomass present

Example 3.2 (Continued)

in the reaction flask (in the example, $M_{sludge} = 0.501$ gVS), gives the specific methanogenic activity of the sludge (mLCH$_4$/gVS.hour). The correspondence of the volume of methane in mass of COD converted into CH$_4$ (COD-CH$_4$) is usually done, as detailed in Chapter 2 (Equations 2.15 and 2.16), so as to enable the SMA to be expressed in terms of gCOD-CH$_4$/gVS·d.

Figure 3.6 shows the methanogenic activity curves for each of the flasks, obtained by calculating the activity for each time interval and not just for the parts where the methane production rate is maximum.

According to Figure 3.6, the maximum activities were approximately 0.50, 0.55, 0.75 and 0.68 gCOD-CH$_4$/gVS·d, for flasks 1, 2, 3 and 4, respectively. In this example, the anaerobic sludge showed its largest activity for a substrate concentration equal to 2.0 gCOD/L (flask 3). This is the specific methanogenic activity of the sludge that should be considered. The most accurate calculation of the activities should be done with the reaches of maximum slope (Figure 3.5), as explained previously.

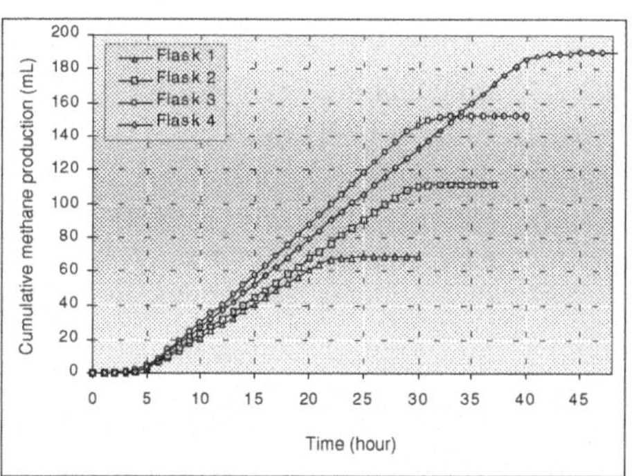

Figure 3.5.   SMA test. Cumulative CH$_4$ production results

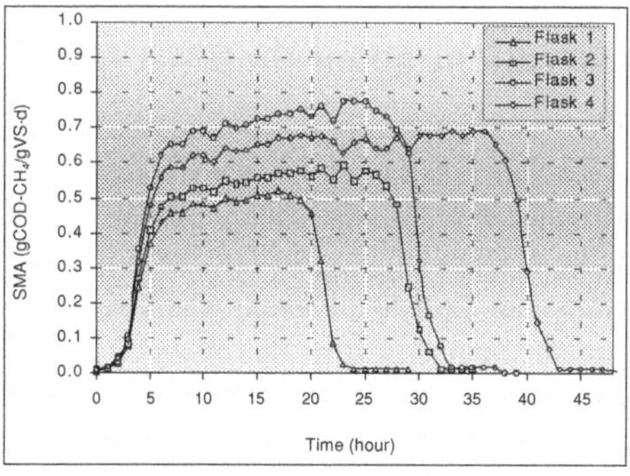

Figure 3.6. SMA test. Methanogenic activity results

Example 3.2 (Continued)

- Determination of the amount of substrate converted into methane:
  According to the curves of Figure 3.5, the total CH_4 production, at the end
  of the test for each of the flasks, was:
  - flask 1: $V_{CH_4} \cong 70$ mL
  - flask 2: $V_{CH_4} \cong 112$ mL
  - flask 3: $V_{CH_4} \cong 152$ mL
  - flask 4: $V_{CH_4} \cong 190$ mL
- Determination of the theoretical methane production, from the amount of
  substrate (gCOD) added to each flask:

  According to Equations 2.15 and 2.16 (Chapter 2):

$$K(t) = (P \cdot K)/[R \cdot (273 + T)] = (1 \times 64)/[0.08206 \times (273 + 30)]$$

$$= 2.57 \text{ gCOD/L}$$

$$V_{CH_4} = COD\text{–}CH_4/K(t) =$$

  - flask 1: 2 mL $\times$ 100 mgCOD/mL = 200 mgCOD $\Rightarrow$ VCH$_4$ =
    200 mgCOD/2.57 mgCOD/mL = 77.8 mL
  - flask 2: 3 mL $\times$ 100 mgCOD/mL = 300 mgCOD $\Rightarrow$ VCH$_4$ =
    300 mgCOD/2.57 mgCOD/mL = 116.7 mL
  - flask 3: 4 mL $\times$ 100 mgCOD/mL = 400 mgCOD $\Rightarrow$ VCH$_4$ =
    400 mgCOD/2.57 mgCOD/mL = 155.6 mL
  - flask 4: 5 mL $\times$ 100 mgCOD/mL = 500 mgCOD $\Rightarrow$ VCH$_4$ =
    500 mgCOD/2.57 mgCOD/mL = 194.6 mL
- Determination of the percentage substrate converted into methane:
  - flask 1: 70 mL/77.8 mL = 90%
  - flask 2: 112 mL/116.7 mL = 96%
  - flask 3: 152 mL/155.6 mL = 98%
  - flask 4: 190 mL/194.6 mL = 98%

3.4.4 Final considerations about the SMA test

Although the SMA test constitutes a very useful tool, the results should still be used with caution, as there is no accepted international standard as yet. The efforts of the *IWA Task Group on anaerobic biodegradability and activity tests* in establishing such standard should be acknowledged. So far, the different methodologies and experimental conditions can lead to different SMA results, which are difficult to be compared amongst themselves. In this respect, it is understood that the results obtained with the test reflect much more the relative specific methanogenic activities, and not the absolute ones. However, even if the results are relative for certain test conditions, they are very important for the follow-up and evaluation of anaerobic reactors.

4

Anaerobic treatment systems

4.1 PRELIMINARIES

The essence of biological wastewater treatment processes resides in the capacity of the microorganisms involved to use the biodegradable organic compounds and transform them into by-products that can be removed from the treatment system. The by-products formed can be in solid (biological sludge), liquid (water) or gaseous (carbon dioxide, methane, etc.) form. In any process used, aerobic or anaerobic, the capacity for using the organic compounds will depend on the microbial activity of the biomass present in the system.

Until recently, the use of anaerobic processes for the treatment of liquid effluents was considered uneconomical and problematic. The reduced growth rate of the anaerobic biomass, especially the methanogenic Archaea, makes the control of the process delicate, since the recovery of the system is very slow when the anaerobic biomass is exposed to adverse environmental conditions.

With the expansion of research in the area of anaerobic treatment, "high-rate systems" have been developed. Essentially, these are characterised by their ability to retain large amounts of high-activity biomass, even with the application of low hydraulic detention times. Thus, a high solids retention time is maintained, even with the application of high hydraulic loads to the system. The result is compact reactors with volumes inferior to conventional anaerobic digesters, however maintaining the high degree of sludge stabilisation. In this chapter, the main anaerobic systems used for wastewater treatment are described. For convenience, they are classified into two large groups, as shown below:

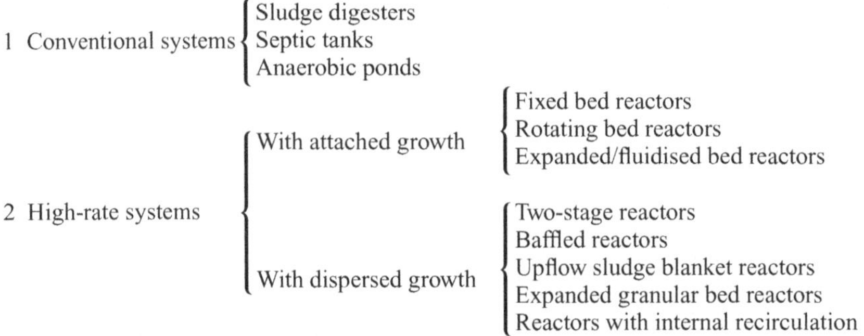

4.2 CONVENTIONAL SYSTEMS

4.2.1 Preliminaries

In this chapter, the designation *conventional systems* is used to classify reactors that are operated with low volumetric organic loads, as they do not have retention mechanisms for large quantities of high-activity biomass. Obviously, a well-defined separation line does not exist between the conventional and the high-rate systems. The examples presented here are only for the purpose of classifying some types of reactors, based on the main aspects that differentiate them from high-rate reactors, which are:

- Absence of solids retention mechanisms in the system: as discussed in Chapter 3, biomass retention in anaerobic systems is improved in a significant way through mechanisms that favour the immobilisation of the microorganisms inside the digestion compartment, as attachment and granulation. The absence of such mechanisms hinders the retention of great amounts of biomass in the treatment system.
- Long hydraulic detention times and low volumetric loads: the absence of solids retention mechanisms in the system implies the need for the conventional reactors to be designed and operated with long hydraulic detention times, to guarantee that the biomass will stay in the system long enough for its growth.
- Low volumetric loads: the design of reactors with long hydraulic detention times implies having tanks with large volumes and, as a result, low volumetric loads applied to the system ($kgCOD/m^3$reactor·d or $kgVS/m^3$reactor·d).

From the following discussion, it will become clear that some aspects that are used to classify conventional systems can be found in a more or less pronounced way in a certain reactor type. It can be inferred that conventional systems are evolving towards high-rate systems.

4.2.2 Anaerobic sludge digesters

Conventional digesters are mainly used for the stabilisation of primary and secondary *sludge*, originating from sewage treatment, and for the treatment of *industrial effluents* with a high concentration of suspended solids. They usually consist of covered circular or egg-shaped tanks of reinforced concrete. The bottom walls are usually inclined, so as to favour the sedimentation and removal of the most concentrated solids. The covering of the reactor can be fixed or floating (mobile).

Since conventional digesters are preferably used for the stabilisation of wastes with a high concentration of particulate material, the hydrolysis of these solids can become the limiting stage of the anaerobic digestion process. The hydrolysis rate, in turn, is affected by several factors, such as: (i) temperature; (ii) residence time; (iii) substrate composition and (iv) particle size.

Thus, with the aim to optimise the hydrolysis of the particulate material, conventional digesters may be heated up, with operation temperatures usually ranging from 25 to 35 °C. The hydrolysis phase evolves very slowly when the digesters are operated at temperatures below 20 °C.

As the conventional digesters do not have specific means for biomass retention in the system, the hydraulic detention time should be long enough to guarantee the permanence and multiplication of the microorganisms in the system, while enabling all the phases of the anaerobic digestion to be processed appropriately.

Depending on the existence of mixing devices and on the number of stages, three main digester configurations have been applied:

- low-rate anaerobic sludge digester
- one-stage high-rate anaerobic sludge digester
- two-stage high-rate anaerobic sludge digester

(a) Low-rate anaerobic sludge digester

The low-rate digester does not have mixing devices and usually comprises a single tank, where the digestion, sludge thickening and supernatant formation occur simultaneously. Raw sludge is added to the part of the digester where the sludge is undergoing active digestion and the biogas is being released. With the upflow movement of the biogas, particles of sludge and other flotation materials are taken to the surface, forming a scum layer. As a result of the digestion, the sludge stratifies below the scum layer, and four different zones are formed inside the reactor, as characterised (see Figure 4.1): scum zone, supernatant zone, active digestion zone and stabilised sludge zone.

The supernatant and stabilised sludge are periodically removed from the digester. Because of the sludge stratification and the absence of mixing, no more than 50% of the digester volume are actually used in the digestion process, with large reactor volumes being required to achieve good sludge stabilisation. In view of these limitations, low-rate digesters are mainly used in small treatment plants.

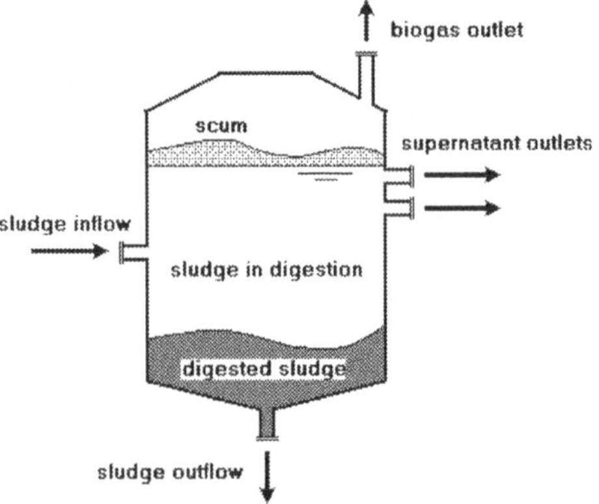

Figure 4.1.   Schematic representation of a low-rate anaerobic sludge digester

(b) One-stage high-rate anaerobic sludge digester

The one-stage high-rate digester incorporates supplemental heating and mixing mechanisms, besides being operated at uniform feeding rates and with the previous thickening of the raw sludge, to guarantee more uniform conditions in the whole digester. As a result, the tank volume can be reduced and the stability of the process is improved. Figure 4.2 presents a schematic representation of a one-stage high-rate digester.

The solids retention times recommended for the design of complete-mix digesters are illustrated in Figure 4.3, and the high dependence of these in relation to the operational temperature of the digester can be observed. When sizing the reactor, the hydraulic detention time shall be equal to the solids retention time, as the system does not have a solids retention mechanism.

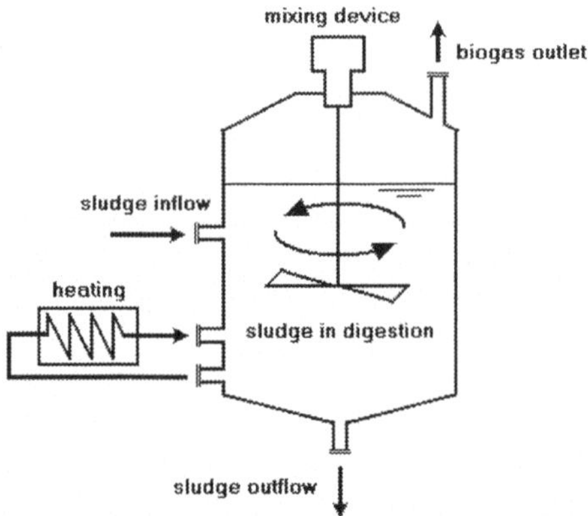

Figure 4.2.   Schematic representation of a one-stage high-rate anaerobic sludge digester

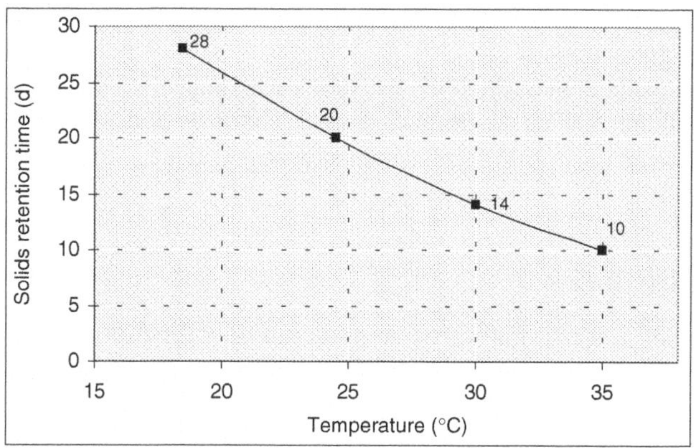

Figure 4.3.   Design recommendations for completely mixed anaerobic digesters (adapted from Metcalf and Eddy, 1991)

Different techniques such as gas recirculation, sludge recirculation or mechanical mixers of various configurations can be used to obtain the mixture of the sludge inside the digester.

(c) Two-stage high-rate anaerobic sludge digester

Basically, the two-stage digester consists in the incorporation of a second tank, operating in series with a high-rate primary digester, as illustrated in Figure 4.4. In this configuration, the first tank is used for the digestion of the sludge, and may therefore be equipped with heating and mixing devices. The second tank is used

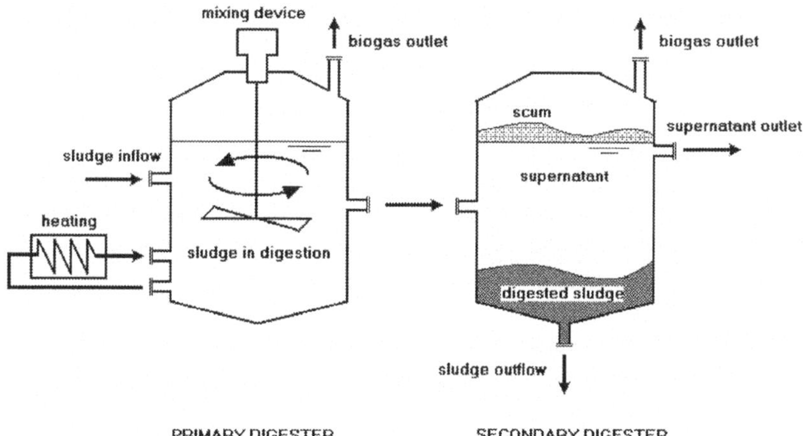

Figure 4.4.   Schematic representation of a two-stage high-rate anaerobic sludge digester

for the storage and thickening of the digested sludge, leading to the formation of a clarified supernatant.

There are situations in which the two tanks are designed in an identical way, so that either can be used as the primary digester. In other situations, the secondary digester can be an open tank, a tank without heating, or even a sludge pond (Metcalf and Eddy, 1991).

4.2.3 Septic tank

The septic tank is a unit that carries out the multiple functions of sedimentation and removal of floatable materials, besides acting as a low-rate digester without mixing and heating capabilities. Septic tanks were conceived around 1860, based on the pioneering work of Mouras, in France. They are still extensively used all over the world and constitute one of the main alternatives for the primary treatment of sewage from residences and small areas that are not served by sewerage networks. The operation of septic tanks can be described as follows:

- The settleable solids present in the influent sewage go to the bottom of the tank and form a sludge layer.
- The oils, grease and other lighter materials present in the influent sewage float on the surface of the tank, forming a scum layer.
- The sewage, free from the settled and floated material, flows between the sludge and scum layers and leaves the septic tank at the opposite end, from where it is directed to a post-treatment unit or to final disposal.
- The organic matter kept at the bottom of the tank undergoes facultative and anaerobic decomposition, and is converted into gaseous compounds such as CO_2, CH_4 and H_2S. Although H_2S is produced in septic tanks, odour problems are not usually observed as it combines with metals accumulated in the sludge and forms insoluble metallic sulfides.
- The anaerobic decomposition provides a continuous reduction of the sludge volume deposited at the bottom of the tank. There is always an accumulation during the months of operation of the septic tank and consequently the sludge and scum accumulation reduces the net volume of the tank, which demands periodic removal of these materials.

To optimise the retention of settleable and floatable solids inside the tank, the tank is usually equipped with internal baffles close to the inlet and outlet points. Multiple compartments are also used with the purpose of reducing the amount of solids in the effluent, although single-chamber tanks are more commonly used, as illustrated in Figure 4.5.

Improvement of the septic tank can be achieved by imposing an upward flow and gas/solid/liquid separation at the top, as in the so-called UASB septic tank (van Lier *et al.*, 2002). This system configuration differs from the conventional septic tank by the upflow mode, which allows a better mixing between the influent and the biomass present at the bottom of the tank, resulting in improved biological conversion of dissolved components. In addition, the upward flow and the gas/solid/liquid

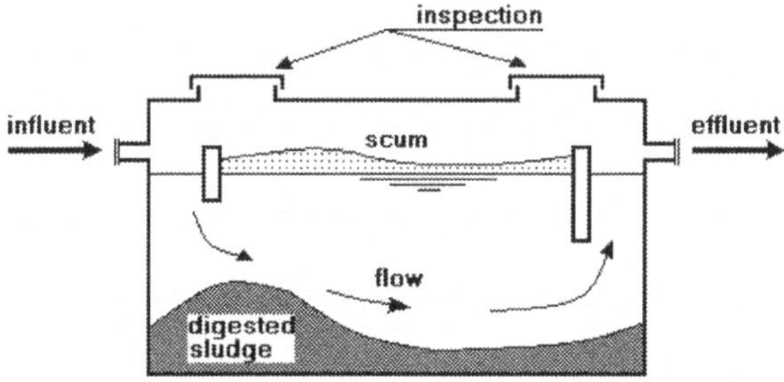

Figure 4.5.   Schematic representation of a single-chamber septic tank

separator enhance the physical removal of suspended solids. The UASB septic tank differs from the conventional UASB reactor (see Section 4.3.3) mainly in relation to sludge accumulation. In the case of UASB septic tank, sludge needs to be removed only once in 1 or 2 years, depending on the design of the reactor.

4.2.4 Anaerobic pond

Anaerobic ponds constitute a very appropriate alternative for sewage treatment in warm-climate regions, and they are usually combined with facultative ponds. They are also frequently used for the treatment of wastewaters with a high concentration of organic matter, such as those from slaughterhouses, dairies, breweries, etc. Figure 4.6 illustrates a typical anaerobic pond.

Owing to the large dimensions and the long hydraulic detention times, anaerobic ponds can be classified as low volumetric organic load reactors. In their typical configuration, the operation of the anaerobic ponds is very similar to that of septic tanks and uses the same basic removal mechanisms described in the previous section. However, the dimensions of the anaerobic ponds are superior to those of the septic tanks, which gives them some different characteristics:

- Because of the great volumes and high depths, there is no need for the systematic removal of the sludge deposited at the bottom of the anaerobic ponds, and cleaning is expected to be required at intervals of a few years.
- Because they are open reactors, and also because of the large areas occupied, there is always the possibility of release of bad odours and proliferation of insects, which requires great care to be taken when choosing their location.

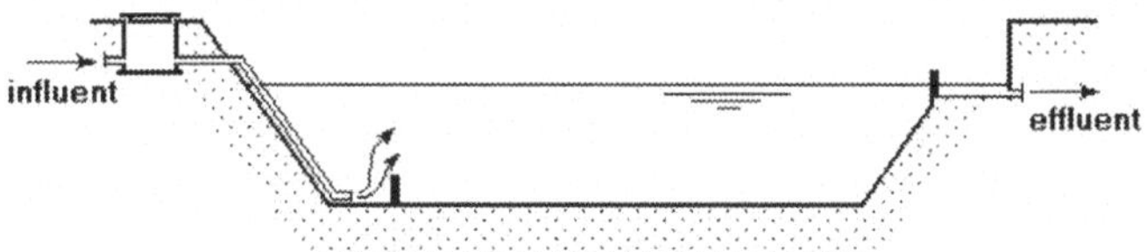

Figure 4.6.   Schematic representation of an anaerobic pond

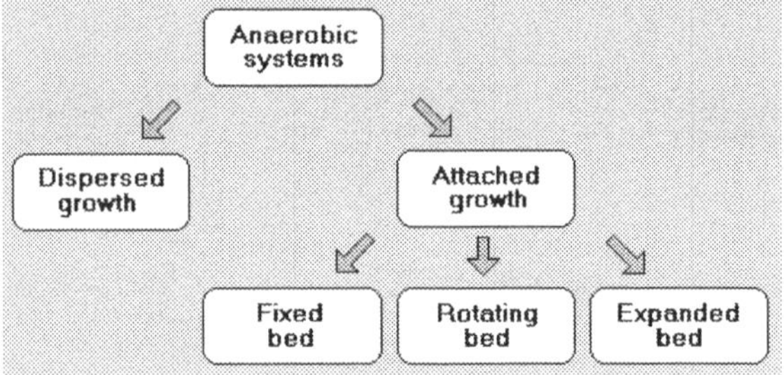

Figure 4.7.   Classification of the anaerobic systems

The main design criteria are based on a volumetric organic load ($kgBOD/m^3 \cdot d$). For domestic sewage, this usually leads to detention times in the order of 3 to 6 days.

Even though the minimum cell residence time of the acetoclastic methanogenic archaea is around 3.3 days, for a temperature of $30\,^{\circ}C$, there has been a recent tendency of reducing the detention times in the anaerobic ponds to around 1 to 2 days. This can be achieved if the retention time of the biomass can be maintained above 3 days, to guarantee the maintenance of a stable bacterial population and an intimate biomass–sewage contact. These conditions can be accomplished through a better distribution of the influent through the bottom of the pond, at several points, aimed at simulating the feeding of UASB reactors (see Section 4.3.3). In this manner, biomass development mechanisms with good settling and activity characteristics are favoured, increasing the solids retention in the system.

4.3 HIGH-RATE SYSTEMS

4.3.1 Preliminaries

As discussed in Chapter 3, anaerobic reactors operated with short hydraulic detention times and long solids retention times need to incorporate biomass retention mechanisms, thereby making up the so-called high-rate systems. Several types of high-rate anaerobic reactors are used for the treatment of sewage and these can be classified into two large groups, according to the type of biomass growth in the system, as illustrated in Figure 4.7.

The concept of dispersed bacterial growth is associated with the presence of free bacterial flocs or granules. On the other hand, the concept of attached bacterial growth requires the development of bacteria joined to an inert support material, leading to the formation of a biological film (biofilm).

4.3.2 Systems with attached bacterial growth

The systems with attached bacterial growth can be divided into *fixed bed*, *rotating bed* and *expanded bed* reactors, as described below (adapted from Stronach *et al.*, 1986).

(a) Fixed bed anaerobic reactors

The more commonly known example of reactors with an attached bacterial growth, in a fixed bed, are the **anaerobic filters**. These are characterised by the presence of a stationary packing material, in which the biological solids can attach to or be kept within the interstices. The mass of microorganisms attached to the support material or kept in their interstices degrades the substrate contained in the sewage flow and, although the biomass is released sporadically, the average residence time of solids in the reactor is usually above 20 days.

The first investigations concerning anaerobic filters date from the end of the 1960s and ever since they have had a growing application in the treatment of different types of industrial and domestic effluents. These filters are usually operated with a vertical flow, upward or downward, with the upflow being more commonly used. In the upflow configuration, the liquid is introduced at the bottom, flows through a filter layer (support medium) and is discharged through the upper part (Figure 4.8). In the downflow configuration, sewage is distributed in the upper part of the filter, above the support medium, and is collected in the lower part of the reactor. Downflow reactors can be used with submerged or non-submerged support medium. Effluent recirculation is more commonly practised in this second configuration (Figure 4.9).

There has been an improvement in the optimisation and efficiency of these systems with the increase of microbiological and biochemical knowledge, which

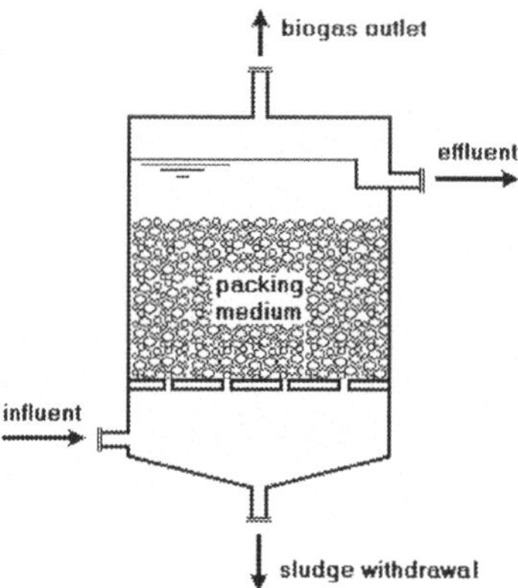

Figure 4.8.   Schematic representation of an upflow anaerobic filter

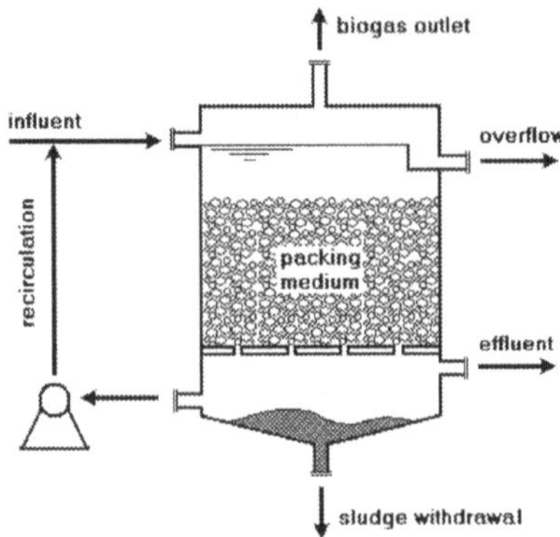

Figure 4.9.   Schematic representation of a downflow anaerobic filter

has enhanced their applicability. It can be verified that the average residence time of the microorganisms in the reactors is very high. This is because they are attached to the support medium, which favours a good treatment process performance.

The most important characteristics of a biological treatment are the solids retention time and the concentration of microorganisms present in the medium. The long solids residence times in the reactors, associated with the short hydraulic detention times, provide the anaerobic filter with a great potential for application to the treatment of low-concentration wastewater. A significant portion of the biomass is found as suspended flocs, which are held in the empty spaces of the support medium (interstitial retention), a fact that caused some researchers to state that the shape of the support material is more important than the type of material employed.

The main disadvantage of anaerobic filters is the accumulation of biomass at the bottom of upflow reactors, where it can lead to blockage or the formation of hydraulic short circuits. In this respect, the downflow filters are more suitable for the treatment of wastes that contain higher concentrations of suspended solids. Further details about the design and operation of anaerobic filters are presented in Chapter 5.

(b) Rotating bed anaerobic reactor

The rotating bed reactor, also called aerobic biodisc, was initially documented in 1928, but it was not until the appearance of plastic materials as effective, light and economical support mediums that the process had a wide application to sewage treatment. In this system, the microorganisms attach to the inert support medium and form a biological film. The support medium, with a sequential disc configuration, is partly or totally submerged and rotates slowly around a horizontal axis in a tank through which the sewage flows.

The anaerobic biodisc was developed by Friedman and Tait (1980). The system configuration is similar to that of the aerobic biodisc (Figure 4.10), except that the tank is covered to avoid contact with air. The submergence of the discs is

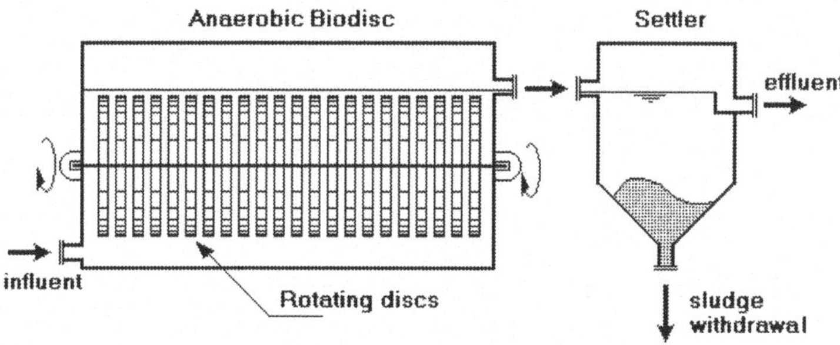

Figure 4.10.   Schematic representation of an anaerobic biodisc

also usually larger than that in the aerobic systems, as the transfer of oxygen is not required. The θ_c/t relation (solids retention time/hydraulic detention time) is very high and blocking should not occur in the system, since the rotation speed of the discs is such that the shearing forces promote the removal of the excess biomass kept between the discs. However, care should be taken in the transfer of results obtained in the laboratory to full-scale plants (scale-up), as the rotation speed substantially increases with the increase of the disc diameter. In high rotation speed conditions, the shearing forces can prevent biomass attachment.

(c) Expanded bed anaerobic reactors

The development of the expanded and fluidised bed anaerobic reactors practically eliminated the problems of the limitation of substrate diffusion, usually inherent to the stationary bed processes. In the expanded and fluidised bed processes the biomass grows into reduced thickness films, attached to small sized particles, in contrast to the stationary bed processes, in which the biofilm has considerably larger thickness and is attached to a support medium also of larger dimensions. The expansion and fluidisation of the medium reduces or eliminates blockage problems, besides increasing the biomass retention and its contact with the substrate, thereby allowing significant reductions in the hydraulic detention times in the reactors. Although the distinction between expansion and fluidisation is frequently not clearly defined, two main systems can be characterised.

Expanded bed anaerobic reactor. The process of attached growth and expanded bed was developed by Jewell (1981), as an extension of the existent anaerobic processes. The expanded bed reactors consist of a cylindrical structure, packed with inert support particles to about 10% of its volume. Several types of materials have been used as support mediums, including sand, gravel, coal, PVC, resins, etc. These support particles, with diameters in the order of 0.3 to 3.0 mm, are slightly larger than those used in fluidised bed reactors. The biofilm grows attached to the particles, which are expanded by the upward velocity of the liquid, increased by the high rate of recirculation applied. The expansion of the bed is maintained

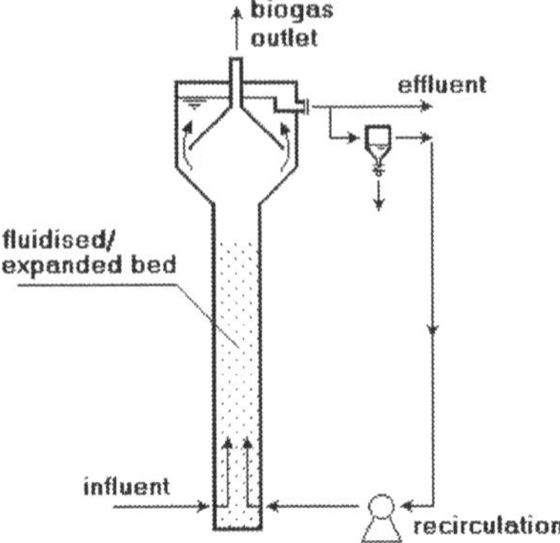

Figure 4.11.   Schematic representation of an expanded/fluidised bed reactor

at a level required for each support particle to preserve its relative position to each of the other particles inside the bed. The expansion of the bed is usually maintained between 10 and 20%. The attached growth and expanded bed reactor was considered the first anaerobic process capable of treating diluted sewage at room temperature (Jewell, 1981). In fact, the system has proved to be very efficient in treating very low concentration sewage (in the range of 150 to 600 mgCOD/L), with minimum hydraulic detention times (in the order of 30 to 60 minutes). In these conditions, COD removal efficiencies of about 60 to 70% can be obtained. The formation of a high-activity biomass, with a concentration in the order of 30 gVSS/L, and the retention and filtration of fine inert particles are the reasons for the high-quality effluent in terms of COD and suspended solids.

Fluidised bed anaerobic reactor. The operating principles of the fluidised bed reactor (Figure 4.11) are basically the same as those of the expanded bed reactor, except for the size of the particles of the support medium and the expansion rates. In this case, the upward velocity of the liquid should be sufficiently high to fluidise the bed until it reaches the point at which the gravitational force is equalled by the upward drag force. A high recirculation rate is required and, as a result, each independent particle does not maintain a fixed position inside the bed. The expansion of very fine particles (0.5 to 0.7 mm) guarantees a very large surface area for the growth of a uniform biofilm around each particle. The expansion degree usually varies between 30 and 100%. Volumetric loads as high as 20 to 30 kgCOD/m^3·d have been reported using soluble wastes of medium and high concentrations, with COD removal efficiencies between 70 and 90%.

4.3.3 Systems with dispersed bacterial growth

The efficiency of the systems with dispersed bacterial growth depends largely on the capacity of the biomass to form flocs and settle. Included among the processes with dispersed bacterial growth are the two-stage reactors, baffled reactors and the

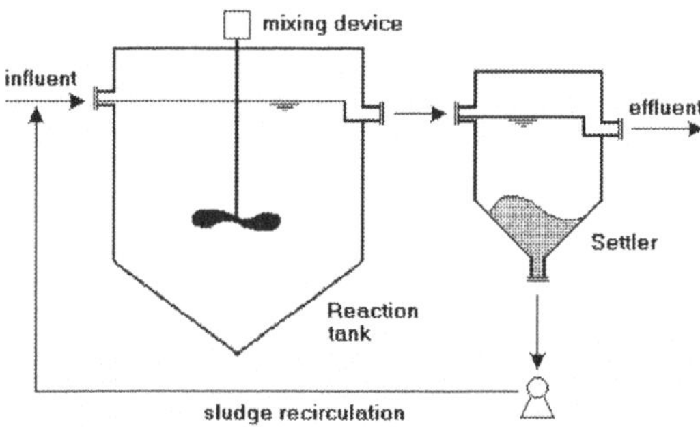

Figure 4.12.   Schematic representation of a two-stage reactor

upflow sludge blanket reactors and their variants (expanded granular sludge bed and anaerobic reactor with internal recirculation).

(a) Two-stage anaerobic reactor

The two-stage anaerobic reactor (anaerobic contact process) (Figure 4.12) was developed in the 1950s for the treatment of concentrated industrial wastewater. The system involves the use of a complete-mix tank (anaerobic reactor) followed by a device for the separation and the return of solids. Conceptually, the two-stage reactor is similar to the aerobic activated sludge system. The essence of the two-stage process is that the biomass that is flocculated in the reactor, along with the undigested influent solids that are taken out of the system, is retained through a solids separation device and returned to the first stage reactor where it is mixed with the influent wastewater. The practical difficulty of the two-stage process is the separation and concentration of the effluent solids, as the presence of gas-producing particles leads the biomass flocs to float instead of settling. Several methods have been used or recommended to eliminate these problems, through sedimentation, chemical flocculation, vacuum degasification, flotation and centrifugation, thermal shock, filter membrane, etc.

(b) Baffled anaerobic reactor

The baffled reactor (Figure 4.13) resembles a septic tank with multiple chambers in series and with a more effective feeding device to the chambers. To obtain this configuration, the reactor is equipped with vertical baffles that force the liquid to make a sequential downflow and upflow movement, to guarantee a larger contact of the wastewater with the biomass present at the bottom of the unit. According to Campos (1994), this reactor presents several of the main advantages of the UASB reactors and could be built without the gas separator, therefore with smaller depths, which facilitates its burying, thus representing a reduction in construction costs. However, the project characteristics are not always adequate to guarantee good operational conditions in larger size units. For instance, an excessive loss of solids, in the case of great variations and excessive peaks of the influent flow, may

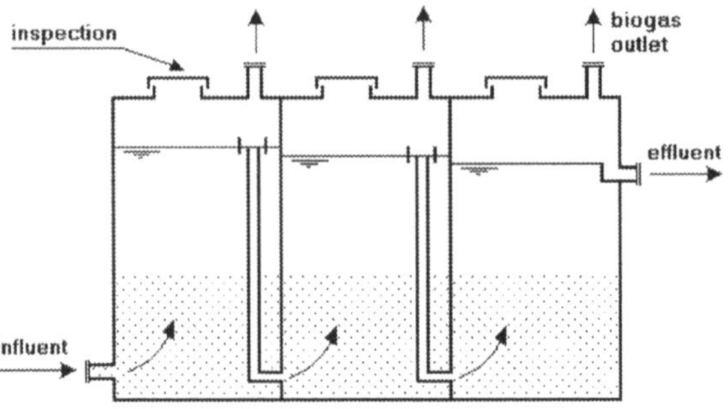

Figure 4.13.   Schematic representation of a baffled reactor

occur in this type of reactor, as the system does not have auxiliary mechanisms for biomass retention.

(c) Upflow anaerobic sludge blanket reactor

The upflow anaerobic sludge blanket (UASB) reactor was developed by Lettinga and co-workers, being initially largely applied in Holland. The process essentially consists of an upflow of wastewater through a dense sludge bed with high microbial activity. The solids profile in the reactor varies from very dense and granular particles with good settleability close to the bottom (*sludge bed*) to a more dispersed and light sludge close to the top of the reactor (*sludge blanket*).

Conversion of organic matter takes place in all reaction areas (bed and sludge blanket), and the mixing of the system is promoted by the upward flow of wastewater and gas bubbles. The wastewater enters at the bottom and the effluent leaves the reactor through an internal settling tank in the upper part of the reactor. A gas and solids separation device located below the settling tank guarantees optimal conditions for sedimentation of the particles that stray from the sludge blanket, allowing them to return to the digestion compartment instead of leaving the system. Although part of the lightest particles is lost together with the effluent, the average solids retention time in the reactor is maintained sufficiently high to sustain the growth of a dense mass of methane-forming microorganisms, in spite of the reduced hydraulic detention time.

One of the fundamental principles of the process is its ability to develop a high-activity biomass. This biomass can be in the form of flocs or granules (1 to 5 mm). The cultivation of a good-quality anaerobic sludge is achieved through a careful start-up of the process, during which the artificial selection of the biomass is imposed, allowing the lightest poor-quality sludge to be washed out of the system while retaining the good-quality sludge. The heaviest sludge usually grows close to the bottom of the reactor, presenting a total solids concentration in the order of 40 to 100 gTS/L. Normally mechanical mixing devices are not used, as they seem to have an adverse effect on the aggregation of the sludge and, consequently, on the formation of granules.

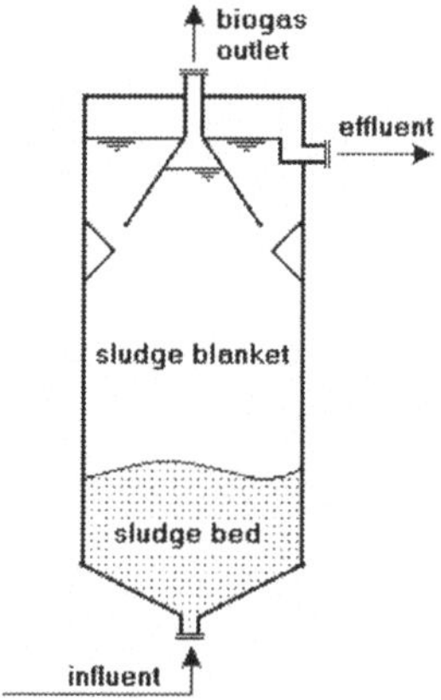

Figure 4.14.   Schematic representation of an UASB reactor

The second fundamental principle of the process is the presence of a gas and solids separation device, which is located in the upper part of the reactor. The main purpose of this device is the separation of the gases contained in the liquid mixture, so that a zone favouring sedimentation is created in the upper part of the reactor.

The design of UASB reactors (Figure 4.14) is very simple and does not require the installation of any sophisticated device or packing medium for biomass attachment and retention. The process was initially developed for the treatment of concentrated wastewater, with very good results. However, similarly to the expanded bed process, in warm-climate regions, UASB reactors have also been applied for the treatment of low-concentration wastewater (domestic sewage) with very good results. As a consequence, UASB reactors are currently one of the preferred alternatives for sewage treatment in these regions. More details about the design and operation of UASB reactors are given in Chapter 5.

(d) Expanded granular sludge bed anaerobic reactor

The expanded granular sludge bed (EGSB) anaerobic reactor (Figure 4.15) greatly resembles the UASB reactor, except in respect to the sludge type and the expansion degree of the sludge bed. Mainly granular-type sludge is retained in the EGSB reactor and is maintained expanded because of the high hydraulic rates applied to the system. This condition intensifies the hydraulic mixing in the reactor and makes a better biomass–substrate contact. The high surface velocities of the liquid in the reactor (in the order of 5 to 10 m/hour) are achieved through the application of a high effluent recirculation rate, combined with the use of reactors with a high height/diameter ratio, around 20 or more (Kato, 1994; Lettinga, 1995). In contrast, in the UASB reactors, the sludge bed remains somewhat static, since the surface velocities of the liquid are usually lower, in the order of 0.5 to 1.5 m/hour.

Regarding the applicability of EGSB reactors, these are mainly intended for the treatment of soluble effluents, as the high surface velocities of the liquid inside

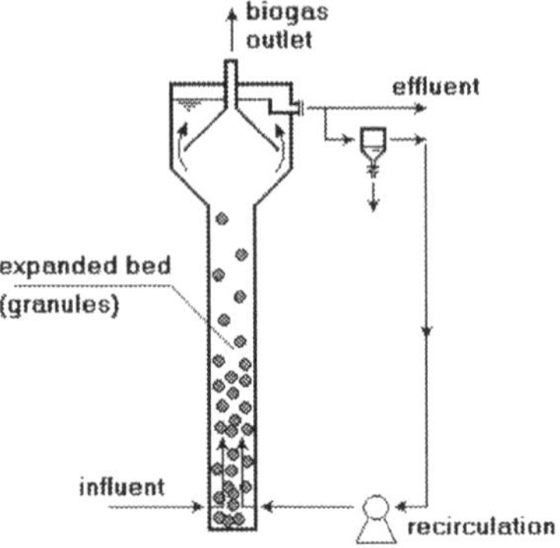

Figure 4.15.   Schematic representation of an expanded granular bed reactor

the reactor do not enable the efficient removal of particulate organic materials. In addition, the excessive presence of suspended solids in the influent can be detrimental to the maintenance of the good characteristics of the granular sludge in the reactor.

As a practical result of the high upward velocities applied to the expanded granular sludge bed reactors, they can be much higher, in the order of 20 m, which results in a significant reduction in the area required. This is particularly interesting in the case of treatment of soluble effluents from industries with little space available. Figure 4.16 illustrates the volumetric organic loads that can be applied to EGSB and UASB reactors considering the treatment of low-concentration soluble wastewater assuming: (i) a granular sludge concentration of 25 gVSS/L; and (ii) 100% acidified effluent (volatile fatty acids).

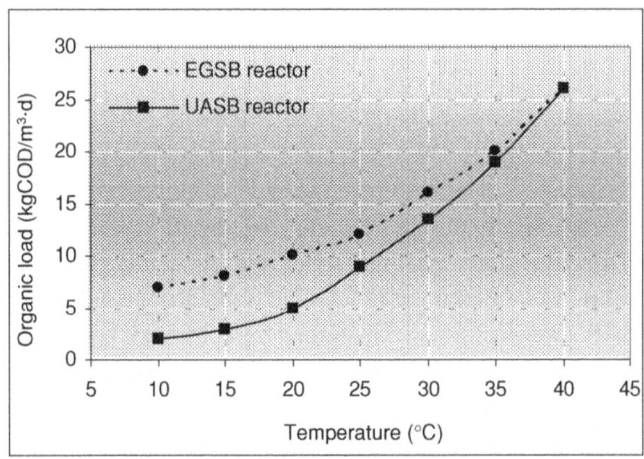

Figure 4.16.   Volumetric organic loads in UASB and EGSB reactors (adapted from Lettinga, 1995)

(e) Anaerobic reactor with internal recirculation

The anaerobic reactor with internal recirculation can be considered a variation of the UASB reactor, and has been developed with the objective of guaranteeing a larger efficiency when submitted to high volumetric organic loads (up to 30 to 40 kgCOD/m^3·d). To allow the application of high loads, it is necessary to have a more efficient gas, solids and liquid separation, as the high turbulence caused by the production of gases hinders the biomass retention in the system.

In the reactor with internal recirculation, the gas, solids and liquid separation is done in two stages:

- In the first stage the separation of the largest portion of the biogas produced in the system occurs, thereby decreasing the turbulence in the upper part of the reactor.
- In the second stage the separation of the solids occurs, which guarantees high biomass retention in the system and a more clarified effluent.

Basically, the reactor with internal recirculation consists of two UASB reactor compartments, one on top of the other, with the first compartment being subjected to high organic loads. This specific task of gas separation in two stages is done in a larger height reactor (16 to 20 m), making the gases collected in the first stage drag the internal mixture (gas, solids and liquid) to the upper part of the reactor (gas lifting effect). After the separation of the gases in the upper part of the reactor, solids and liquids recirculate to the first compartment, which provides high mixing and the contact of the recirculated biomass with the influent wastewater at the base of the reactor (see Figure 4.17).

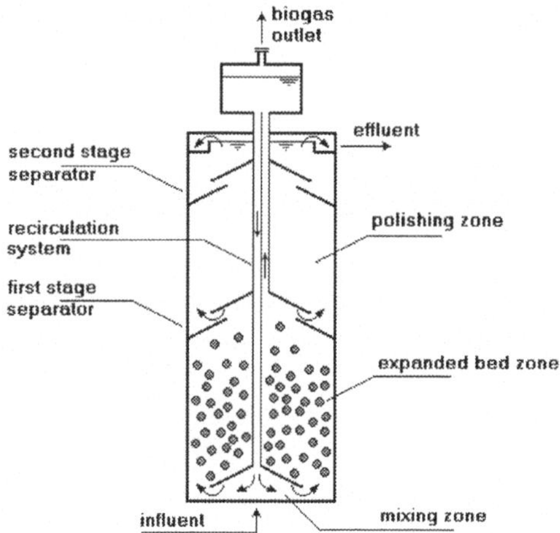

Figure 4.17.   Schematic representation of a reactor with internal recirculation

According to Yspeert et al. (1995), the reactor with internal recirculation incorporates four basic items:

- *Mixing zone*: located at the bottom of the reactor, making possible an effective mixture of the influent wastewater with the biomass and the effluent from the recirculation device. This results in dilution and conditioning of the raw influent waste.
- *Expanded bed zone*: located immediately above the base of the reactor and constitutes the first stage of the reactor. This area contains the high-concentration granular sludge maintained expanded owing to the high upflow velocities caused by the influent, by the recirculation flow and by the biogas produced. The effective contact between the influent waste and the biomass results in a high sludge activity, making possible the application of high organic loads, and in high conversion rates. The high intensity of the biomass mixing in the zone favours the application of this reactor type for the treatment of highly concentrated wastewaters.
- *Polishing zone*: constitutes the second stage of the reactor and is located immediately above the separator of the expanded bed zone. In this area, effective post-treatment and additional biomass retention occur owing to three principal aspects: (i) low applied loads; (ii) high hydraulic detention times; and (iii) proximity to a plug-flow regime. As a result of the almost complete biodegradable COD removal in the expanded bed zone and the collection of gases by the first separator, the turbulence caused by the upward velocity of the liquid in the polishing zone is low.
- *Recirculation system*: comprises a device that makes the internal circulation possible through the gas-lift principle. This condition is created by the difference in the biogas capture between the upflow (gas, solids and liquid flow) and downflow (solids and liquid flow) branches of the recirculation system, without the need for any type of pumping. In studies performed in a pilot reactor of 17 m^3, treating wastes with a concentration of 3,500 mgCOD/L, a recirculation flow approximately 2.5 times the gas flow was obtained.

4.4 COMBINED TREATMENT SYSTEMS

In this chapter, the main anaerobic systems currently used for the treatment of solid and liquid wastes were described and classified, for convenience, into *conventional systems* and *high-rate systems*. There is a consensus that, in most of the applications, the anaerobic systems should be considered a first stage of the treatment, as they are not capable of producing final effluents with very good quality.

Obviously, in some situations, depending on the characteristics of the influent wastewater and the final discharge quality requirements, anaerobic systems can constitute complete treatment, or the first phase (in time) in the implementation of the treatment system along the planning horizon. However, in most of the situations, a combined treatment system has been used to obtain the substantial advantages of the incorporation of an anaerobic system as the first stage, followed by a post-treatment system. In this respect, several post-treatment alternatives have been researched, reported and implemented in the last few years, including both aerobic and anaerobic systems. Virtually all processes capable of treating raw sewage are also capable of acting as post-treatment for the effluent from anaerobic reactors. Post-treatment of anaerobic effluents is covered in Chapter 7.

5

Design of anaerobic reactors

5.1 ANAEROBIC FILTERS

5.1.1 Preliminaries

The first works on anaerobic filters date from the late 1960s and ever since they have had a growing application, representing today an advanced technology for the effective treatment of domestic sewage and a diversity of industrial effluents. The upflow anaerobic filter is basically a contact unit, in which sewage passes through a mass of biological solids contained inside the reactor. The biomass retained in the reactor can be in three different forms:

- thin biofilm layer attached to the surfaces of the packing medium
- dispersed biomass retained in the interstices of the packing medium
- flocs or granules retained in the bottom compartment, below the packed bed

The soluble organic compounds contained in the influent sewage come in contact with the biomass, being diffused through the surfaces of the biofilm or the granular sludge. They are then converted into intermediate and final products, specifically methane and carbon dioxide.

The usual configurations of anaerobic filters are either upflow or downflow. In upflow filters, the packing bed is necessarily submerged. The downflow filters can work either submerged or non-submerged. They are usually covered, but they can be implemented uncovered, when there is no concern with the possible release of bad odours.

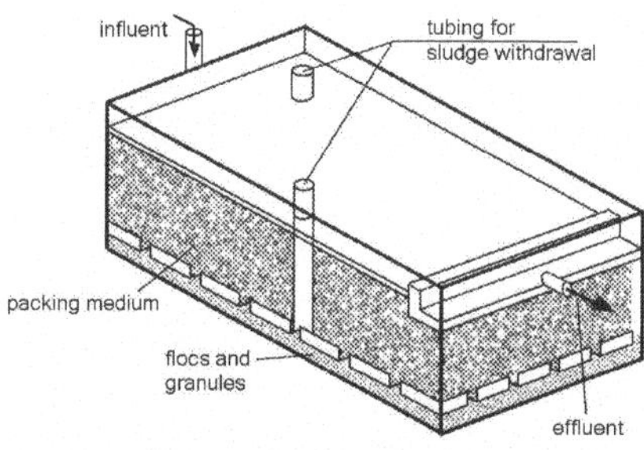

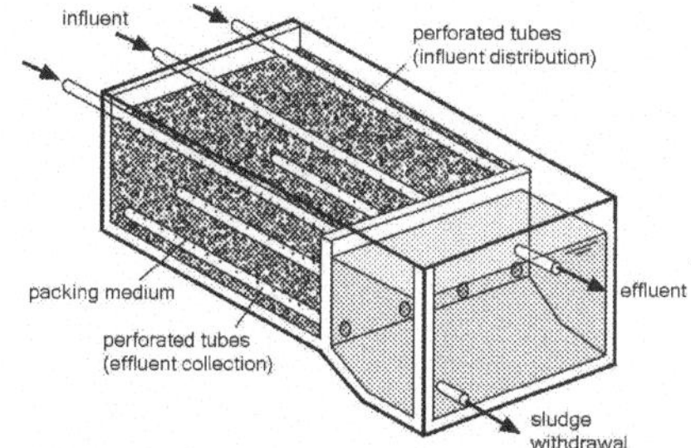

Figure 5.1. Schematic drawing of an upflow anaerobic filter (adapted from Gonçalves et al., 2001)

Figure 5.2. Schematic drawing of a submerged downflow anaerobic filter (adapted from Gonçalves et al., 2001)

Figures 5.1 and 5.2 present schematic drawings of submerged downflow and upflow anaerobic filters, where the main devices that guarantee the proper functioning of the treatment unit can be observed (Gonçalves *et al.*, 2001).

Although anaerobic filters can be used as the main wastewater treatment unit, they are more appropriate for post-treatment (polishing), adding operational safety and stability to the treatment system as a whole.

The effluent from anaerobic filters is usually well clarified and has a relatively low concentration of organic matter, although it is rich in mineral salts. It is very good for land application, not only for infiltration, but also for irrigation with crop production purposes, provided that the concern with pathogenic microorganisms, usually present in large amounts in the effluents from filters that treat domestic sewage, is not disregarded. In these cases, disinfection may become necessary, and the usual existing processes can be applied.

The main limitations of the anaerobic filters result from the risk of bed obstruction (clogging of the interstices) and from the relatively large volume, due to the space occupied by the inert packing material.

Anaerobic filters have been used in different system configurations in Brazil, for the post-treatment of effluents from medium and large anaerobic reactors, as illustrated in Figures 5.3 and 5.4 (Gonçalves *et al.*, 2001).

5.1.2 Physical aspects

(a) Reactor configuration

Anaerobic filters can have several shapes, configurations and dimensions, provided that the flow is well distributed over the bed. In full scale, anaerobic filters usually present either a cylindrical or a rectangular shape. The diameters (or width) of the tanks vary from 6 to 26 m, and their height from 3 to approximately 13 m. The volumes of the reactors vary from 100 to 10,000 m^3. The packing media have been designed to occupy from the total depth of the reactor to approximately 50 to 70% of the height of the tanks. There are different types of plastic packing

Figure 5.3.   Anaerobic filter after upflow anaerobic sludge blanket (UASB) reactor (source: Colombo WWTP, SANEPAR/ Brazil)

Figure 5.4.   Anaerobic filter after UASB reactor (source: Ipatinga WWTP, COPASA, Brazil)

mediums available in the market, ranging from corrugated rings to corrugated plate blocks. The specific surface area of these plastic materials usually ranges from 100 to 200 m^2/m^3. Although some types of packing media are more efficient than others in the retention of biomass, the final choice will depend on the local specific conditions, on economic considerations and on operational factors.

The most recent installations of upflow anaerobic filters have been of the hybrid type, in which there is a zone without packing material, located at the lower part of the reactor, which allows the accumulation of granular sludge. The performance of the hybrid anaerobic filters depends on the contact of the wastewater with the biomass dispersed on the sludge bed and with the biofilm attached to the packing medium. The determination of the amount of packing material to be used in hybrid reactors is still subjective. There is a minimum amount that should be enough to promote some complementary removal of organic matter, and also to help in the retention of biological solids. As recommended by Young (1991), the packed bed should be placed in the upper two-thirds of the height of the reactor, and this medium should not be lower than 2 m. Lower heights should only be adopted from pilot tests or in full-scale systems treating the same type of effluent.

It should be emphasised that the recommendations made by Young (1991) refer mainly to the use of anaerobic filters for treatment of industrial effluents, a situation in which the COD removal occurs throughout the height of the packed bed. In the treatment of more diluted effluents, such as domestic sewage, the removal of organic matter occurs mainly in the lower part of the anaerobic filter (in the bottom compartment and in the beginning of the packed bed), which leads to the use of reduced heights of packing medium.

(b)   Packing medium

The purpose of the packing medium is to retain solids inside the reactor, either by the biofilm formed on the surface of the packing medium or by the retention

Table 5.1. Requirements for packing media of anaerobic filters

Requirement	Objective
■ Be structurally resistant	■ Support their own weight, added to the weight of the biological solids attached to the surface
■ Be biologically and chemically inert	■ Allow no reaction between the bed and the microorganisms
■ Be sufficiently light	■ Avoid the need for expensive, heavy structures, and allow the construction of relatively higher filters, which implies a reduced area necessary for the installation of the system
■ Have a large specific area	■ Allow the attachment of a larger quantity of biological solids
■ Have a high porosity	■ Allow a larger free area available for the accumulation of bacteria and reduce the possibility of clogging
■ Enable the accelerated colonisation of microorganisms	■ Reduce the start-up time of the reactor
■ Present a rough surface and a non-flat format	■ Ensure good attachment and high porosity
■ Have a reduced price	■ Make the process feasible, not only technically, but also economically

Source: Adapted from Pinto and Chernicharo (1996) and Souza (1982), quoted by Carvalho (1994)

of solids in the interstices of the medium or below it. The main purposes of the support layer are as follows:

- to act as a device to separate solids from gases
- to help promote a uniform flow in the reactor
- to improve the contact between the components of the influent wastewater and the biological solids contained in the reactor
- to allow the accumulation of a large amount of biomass, with a consequently increased solids retention time
- to act as a physical barrier to prevent solids from being washed out from the treatment system

Table 5.1 presents the main desirable requirements for packing medium of anaerobic filters.

Several types of materials have been used as packing media in biological reactors, including quartz, ceramic blocks, oysters and mussel shells, limestone, plastic rings, hollow cylinders, PVC modular blocks, granite, polyethylene balls, bamboo, etc.

Recent studies demonstrated the applicability and feasibility of another packing medium alternative for anaerobic filters: blast furnace slag. This material has been used for over 5 years, and no indication of deterioration or bed clogging has been noticed. The samples removed for analyses demonstrated the integrity of the stones and the high attachment capacity of the anaerobic biofilm (Pinto, 1995; Pinto and Chernicharo, 1996).

The clogging of the packing medium has been one of the main concerns of designers and users of anaerobic filters. These problems are more associated with upflow anaerobic filters using stone and crushed stone as packing material. The most modern filters, packed with plastic material, have had no clogging problems, even when the specific surface areas of the packing medium are low, amounting to 100 m^2/m^3. To minimise the clogging effects of the packing medium, cleaning devices should be considered over the height of the filter, to remove the excess solids retained in the filtering medium. The operational aspects are also important to avoid the clogging of the filter, as discussed in Chapter 6.

5.1.3 Hydraulic aspects

(a) Recirculation of effluent

The function and benefits of effluent recirculation in anaerobic filters are not well defined yet. By means of experiments made in laboratories, it has been noticed that the application of recirculation rates of up to 10 times the influent flow provides an improved efficiency to the system. A significantly reduced efficiency was noticed above the recirculation ratio of 10:1.

Recirculation of effluents from either upflow or downflow anaerobic filters is not usually necessary when treating domestic effluents from septic tanks, considering that the concentrations of influent organic matter to the anaerobic filter are not very high (Andrade Neto, 1997).

The recirculation of effluents should not be the first method to lessen the transient conditions of influent loads. High recirculation rates can cause the increase of the upflow velocities, with the consequent loss of biomass.

(b) Upflow velocity

Besides the hydraulic detention time and the effluent recirculation, other hydraulic factors intervening in the process are the upflow velocity and the flow variations. The upflow velocity should be maintained below the limit above which solids are significantly lost in the effluent. In full-scale reactors, the upflow velocity, including the recirculation flow, is usually around 2 m/hour. However, the maximum upflow velocity depends on the density of the suspended solids and on the magnitude of the granulation. The upflow velocity should be maintained low during the start-up of the process, to reduce solids wash out in the effluent. During start-up, effluent recirculation can favour pH control in the reactor, so that the upflow velocities (including the recirculation) do not exceed 0.4 m/hour. The recirculation rates can be gradually increased as the reactor advances to maturity, but upflow velocities higher than 1.0 m/hour can cause an excessive loss of solids.

5.1.4 Performance relationships

Although pilot and laboratory studies contribute to the development of relationships between the several design and operational factors, a general relationship of unrestricted acceptance has not yet been developed to be used in the design of full-scale anaerobic filters.

Young (1991) gathered operational data from several anaerobic filters and established a statistical correlation among them, aiming at the determination of the parameters that influenced the performance of the system. The parameters analysed in the multiple linear regression models included hydraulic detention time, wastewater concentration, surface area of the packing medium, slope of the corrugated plates of the packing medium and volumetric organic load. The statistical studies indicated that the hydraulic detention time was the parameter that had a higher influence on COD removal efficiency in the system, for reactors packed with both synthetic medium and stones. Regarding the corrugated modules, the increased surface area seemed not to influence significantly the efficiency of the system, while the size of the empty spaces and the geometry of the corrugated material did influence the efficiency of the reactors. In addition, the introduction of the slope of the corrugated plates in the linear regression model had a positive impact on the correlation of the analysed data. The general relationship capable of describing the performance of anaerobic filters treating different types of effluents proposed by Young (1991) was

$$E = 100 \times (1 - S_k \times t^{-m}) \tag{5.1}$$

where:
E = efficiency of the system (%)
t = hydraulic detention time (hour)
S_k = coefficient of the system
m = coefficient of the packing medium

It is worth mentioning that this relation is used to estimate the performance of full-scale and laboratory reactors with relative precision, when they use cross-flow synthetic packing medium with a surface area of approximately $100 \ m^2/m^3$. For this situation, the coefficients S_k and m assume values of 1.0 and 0.55, respectively. For stone bed reactors, the value of the coefficient m is approximately 0.40.

Treatment efficiency is also related to temperature by means of the following expression:

$$E_T = 1 - (1 - E_{30}) \theta^{(T-30)} \tag{5.2}$$

where:
E_T = efficiency of the process at temperature T (°C)
E_{30} = efficiency of the process at the temperature of 30 °C
T = operational temperature (°C)
θ = temperature coefficient (1.02 to 1.04)

5.1.5 Design criteria

The use of anaerobic filters for the treatment of domestic wastewater has been intended mainly for the polishing of effluents from septic tanks and UASB reactors. In this serial configuration, the main design considerations are described below.

(a)   Hydraulic detention time

The hydraulic detention time refers to the average time of residence of the liquid inside the filter, calculated by the following expression:

$$\boxed{t = \frac{V}{Q}} \tag{5.3}$$

where:

  t = hydraulic detention time (hour)
  V = volume of the anaerobic filter (m^3)
  Q = average influent flowrate (m^3/d)

In the case of anaerobic filters applied to the post-treatment of effluents from anaerobic reactors, the design criteria and parameters are still very scarce. The result of studies developed by the Brazilian National Research Programme on Basic Sanitation, PROSAB (Gonçalves *et al.*, 2001), using anaerobic filters filled with a stone bed for the polishing of effluents from septic tanks and UASB reactors, showed that they are capable of producing effluents that meet less stringent discharge standards (BOD $\leq$ 60 mg/L, TSS $\leq$ 40 mg/L), when operated under hydraulic detention times ranging from **4 to 10 hours**.

(b)   Temperature

Anaerobic filters can be satisfactorily operated at temperatures ranging from 25 to 38 °C. Usually, the degradation of complex wastewater, whose first stage of the fermentation process is hydrolysis, requires temperatures higher than 25 °C. Otherwise, hydrolysis may become the limiting stage of the process.

In spite of the recommendation that anaerobic filters should be operated within the temperature range from 25 to 38 °C, satisfactory results have been observed for filters operating within the temperature range from 20 to 25 °C (and even lower), especially when applied to the post-treatment of effluents from septic tanks and UASB reactors (Gonçalves *et al.*, 2001) .

(c)   Packing medium height

Based on the Brazilian experience and on studies developed by the Brazilian National Research Programme on Basic Sanitation, PROSAB (Gonçalves *et al.*, 2001) using anaerobic filters filled with a stone bed for the polishing of effluents from septic tanks and UASB reactors, it is recommended for most applications that the packed bed height should be between **0.8 and 3.0 m**. The upper height limit of the packed bed is more appropriate for reactors with lower risk of bed obstruction, which depends mostly on the flow direction, on the type of packing material and on the influent concentrations. A more usual value should amount to approximately **1.5 m**.

(d) Hydraulic loading rate

The hydraulic loading rate refers to the volume of wastewater applied daily per unit area of the filter packing medium, as calculated by Equation 5.4,

$$\boxed{HLR = \frac{Q}{A}} \tag{5.4}$$

where:

HLR = hydraulic loading rate ($m^3/m^2 \cdot d$)

Q = average influent flowrate (m^3/d)

A = surface area of the packing medium (m^2)

The result of studies developed by the Brazilian National Research Programme on Basic Sanitation, PROSAB (Gonçalves *et al.*, 2001), using anaerobic filters filled with a stone bed for the polishing of effluents from septic tanks and UASB reactors, showed that the filters are capable of producing effluents of good quality when operated under surface hydraulic loading rates ranging from **6 to 15 $m^3/m^2 \cdot d$**.

(e) Organic loading rate

The volumetric organic loading rate refers to the load of organic matter applied daily per unit volume of the filter or packing medium, as calculated by Equation 5.5,

$$\boxed{L_v = \frac{Q \times S_0}{V}} \tag{5.5}$$

where:

L_v = volumetric organic loading rate ($kgBOD/m^3 \cdot d$ or $kgCOD/m^3 \cdot d$)

Q = average influent flowrate (m^3/d)

S_0 = influent BOD or COD concentration ($kgBOD/m^3$ or $kgCOD/m^3$)

V = total volume of the filter or volume occupied by the packing medium (m^3)

While anaerobic filters have been designed to support organic loads of up to 16 $kgCOD/m^3 \cdot d$ (considering the total volume), the operational loads do not usually exceed 12 $kgCOD/m^3 \cdot d$, except when the wastewater presents concentrations higher than 12,000 mgCOD/L. This implies the existence of a concentration above which filters are designed based on the organic loading criterion, and below which the design is based on the hydraulic loading criterion. For the treatment of *domestic*

Figure 5.5.   Sewage distribution device at the bottom of an anaerobic filter (Ipatinga WWTP, COPASA, Brazil)

Figure 5.6.   Effluent collection launder on the top of an anaerobic filter (Ipatinga WWTP, COPASA, Brazil)

sewage, the design of anaerobic filters is ruled by the hydraulic detention time parameter.

Studies made by PROSAB indicated that the anaerobic filters are capable of producing good-quality effluents when operated under organic loading rates from **0.15 to 0.50 kgBOD/m^3·d** (total filter volume) and from **0.25 to 0.75 kgBOD/m^3·d** (packed bed volume).

(f)   Effluent distribution and collection systems

A very important aspect of the design of anaerobic filters concerns the detailing of the wastewater inlet and outlet devices, since the efficiency of the treatment system depends substantially on the good distribution of the flow on the packing bed, and this distribution is subject to the correct calculation of the inlet and outlet devices.

In the case of upflow anaerobic filters, one flow distribution tube has been used for **every 2.0 to 4.0 m^2** of filter bottom area. Figures 5.5 and 5.6 show the wastewater distribution device, through perforated tubes, and the effluent collection launder. The details of the bottom compartment and the perforated slab that will sustain the packing bed are shown in these figures.

(g)   Sludge sampling and removal devices

These devices are intended mainly for monitoring the growth and quality of the biomass in the reactor, enabling more control actions over the solids in the system. Thus, the design of anaerobic filters should allow easy means for the sampling and periodical removal of the sludge, by means of appropriate and sufficient devices. At least two sludge samplers should be included, one close to the bottom and the other immediately below the packed bed, to allow the monitoring of the concentration and height of the sludge bed. Additionally, other sludge samplers can be planned over the height of the packed bed (every 0.5 or 1.0 m). These samplers help considerably to plan the discharge of the excess sludge before it can adversely influence through blockage and clogging of the packing medium.

(h) Efficiencies of anaerobic filters

The expected efficiencies for anaerobic filters can be estimated from the performance relationship presented in Equation 5.1. However, as this relation is empirical, having the hydraulic detention time and the characteristics of the packing medium as main dependent variables, its limitations should be recognised. Van Haandel & Lettinga (1994) propose other empirical constants for Equation 5.1, obtained from the fitting of experimental data from different researches on anaerobic filters:

$$E = 100 \times (1 - 0.87 \times t^{-0.50}) \tag{5.6}$$

where:

E = efficiency of the anaerobic filter (%)

t = hydraulic detention time (hour)

0.87 = empirical constant (coefficient of the system)

0.50 = empirical constant (coefficient of the packing medium)

However, van Haandel and Lettinga (1984) emphasise the limitation of Equation 5.6 in two aspects:

- absence of reports about the use of real-scale anaerobic filters treating domestic sewage
- limited number of data used for determination of the empirical constants of Equation 5.6, which showed great deviations amongst themselves.

Pilot-scale research using anaerobic filters as the first treatment unit, preceded only by preliminary treatment devices (fine screening and grit removal), indicated average BOD and COD removal efficiencies ranging between 68 and 79%. These results were obtained for filters treating domestic wastewater, operating with constant flow and hydraulic detention times varying from 6 to 8 hours (Pinto, 1995).

In situations in which the anaerobic filters are used as post-treatment units for effluents from septic tanks and UASB reactors, the BOD removal efficiencies expected for the system as a whole vary from 75 to 85%.

From the efficiency expected for the system, the COD or BOD concentration in the final effluent can be estimated as follows:

$$C_{eff} = S_0 - \frac{E \times S_0}{100} \tag{5.7}$$

where:

C_{eff} = effluent total BOD or COD concentration (mg/L)

S_0 = influent total BOD or COD concentration (mg/L)

E = BOD or COD removal efficiency (%)

Table 5.2. Design criteria for anaerobic filters applied to the post-treatment of effluents from anaerobic reactors

Design criteria/parameter	Range of values, as a function of the flowrate		
	for $Q_{average}$	for $Q_{daily-maximum}$	for $Q_{hourly-maximum}$
Packing medium	Stone	Stone	Stone
Packing bed height (m)	0.8 to 3.0	0.8 to 3.0	0.8 to 3.0
Hydraulic detention time* (hour)	5 to 10	4 to 8	3 to 6
Surface loading rate (m^3/m^2·d)	6 to 10	8 to 12	10 to 15
Organic loading rate (kgBOD/m^3·d)	0.15 to 0.50	0.15 to 0.50	0.15 to 0.50
Organic loading in the packed bed (kgBOD/m^3·d)	0.25 to 0.75	0.25 to 0.75	0.25 to 0.75

* The adoption of the lower limits of HDT for the design of anaerobic filters requires special care regarding the type of packing medium, the presence of TSS in the influent and the height of the packing bed. Besides that, the operational routine will demand a higher sludge discharge frequency, to avoid clogging problems.
Source: Gonçalves *et al.* (2001)

(i)   Summary of design criteria

A summary of the main criteria and parameters for the design of anaerobic filters, applied to the post-treatment of effluents from anaerobic reactors, as covered in the previous items, is presented in Table 5.2.

Example 5.1

Design an anaerobic filter for the post-treatment of effluents generated in a UASB reactor, with the following design elements being known:

Data:

- Population: P = 20,000 inhabitants
- Average influent flowrate: $Q_{av} = 3{,}000$ m^3/d
- Maximum daily influent flowrate: $Q_{max-d} = 3{,}600$ m^3/d
- Maximum hourly influent flowrate: $Q_{max-h} = 5{,}400$ m^3/d
- Influent organic load to the UASB reactor: $L_{0-UASB} = 1{,}000$ kgBOD/d
- Average influent BOD concentration to the UASB reactor: $S_{0-UASB} = 333$ mg/L
- BOD removal efficiency expected for the UASB reactor: 70%
- Influent organic load to the anaerobic filter: $L_{0-AF} = 300$ kgBOD/d
- Average influent BOD concentration to the anaerobic filter: $S_{0-AF} = 100$ mg/L

Solution:

(a)   Adoption of a hydraulic detention time (t)

According to Table 5.2, the anaerobic filters should be designed with HDT between 3 and 10 hours. Value adopted: t = 8 hours (for average flowrate)

Example 5.1 (Continued)

(b)   Calculation of the volume of the filter, according to Equation 5.3 (V)

$$V = (Q \times t) = [(3{,}000 \text{ m}^3/\text{d})/(24\text{hours}/\text{d})] \times 8 \text{ hours} = 1{,}000 \text{ m}^3$$

(c)   Adopt depth for the packed bed and for the filter:

According to Table 5.2, the anaerobic filters should be designed with packed bed heights between 0.80 and 3.00 m. Value adopted for the packed bed: $h_1 = 1.50$ m

The height of the bottom compartment (h_2) and free depth to the effluent collection launder (h_3) should also be defined. Values adopted: $h_2 = 0.60$ m and $h_3 = 0.30$ m.

The total resulting depth for the filter will be:

$$H = h_1 + h_2 + h_3 = 1.50 + 0.60 + 0.30 = 2.40 \text{ m}$$

(d)   Calculation of the area of the anaerobic filter (A)

$$A = V/H = (1{,}000 \text{ m}^3)/(2.40 \text{ m}) = 416.7 \text{ m}^2$$

(e)   Calculation of the volume of the packed bed (V_{pb})

$$V_{pb} = A \times h_1 = 416.7 \text{ m}^2 \times 1.50 \text{ m} = 625.1 \text{ m}^3$$

(f)   Verification of the hydraulic loading rate (HLR), according to Equation 5.4

For average flowrate: $HLR_1 = Q_{av}/A = (3{,}000 \text{ m}^3/\text{d})/(416.7 \text{ m}^2) = 7.2 \text{ m}^3/\text{m}^2{\cdot}\text{d}$

For maximum daily flowrate: $HLR_2 = Q_{max\text{-}d}/A = (3{,}600 \text{ m}^3/\text{d})/(416.7 \text{ m}^2) = 8.6 \text{ m}^3/\text{m}^2{\cdot}\text{d}$

For maximum hourly flowrate: $HLR_3 = Q_{max\text{-}h}/A = (5{,}400 \text{ m}^3/\text{d})/(416.7 \text{ m}^2) = 13.0 \text{ m}^3/\text{m}^2{\cdot}\text{d}$

According to Table 5.2, it is verified that the surface hydraulic loading rate values are within the recommended ranges for the three flow conditions applied.

(g)   Verification of the average organic loading rate applied to the anaerobic filter and to the packed bed (L_v), according to Equation 5.5

$$L_{v1} = (Q \times S_0)/V = [(3{,}000 \text{ m}^3/\text{d}) \times (0.100 \text{ kgBOD}/\text{m}^3)]/(1{,}000\text{m}^3)$$

$$= 0.30 \text{ kgBOD}/\text{m}^3{\cdot}\text{d}$$

Example 5.1 (*Continued*)

$$L_{v2} = (Q \times S_0)/V_{pb} = [(3{,}000 \ m^3/d) \times (0.100 \ kgBOD/m^3)]/(625.1 \ m^3)$$

$$= 0.48^* \ kgBOD/m^3{\cdot}d$$

(*) In practice, it is noticed that a large part of the influent organic load is removed in the lower part (bottom compartment) of the anaerobic filter, which makes the volumetric organic loads applied to the packed bed much lower.

(h)   Determination of the filter dimensions

Adopt 2 square section filters, each with an area of 208.8 m^2 (14.45 m $\times$ 14.45 m)

(i)   Estimation of the efficiency of the anaerobic filter (E), according to Equation 5.6:

$$E = 100 \times (1 - 0.87 \times t^{-0.50}) = 100 \times (1 - 0.87 \times 8^{-0.50}) = 69\%$$

(j)   Estimation of the BOD concentration in the final effluent (equation 5.7):

$$BOD_{eff} = S_0 - (E{\cdot}S_0)/100 = 100 - (69\% \times 100)/100 = 31 \ mg/L$$

5.2 UPFLOW ANAEROBIC SLUDGE BLANKET REACTORS

5.2.1 Preliminaries

The use of UASB reactors for the treatment of domestic sewage is already a reality in tropical countries, especially in Brazil, Colombia and India. The successful experience in these countries is a strong indication of the potential of this type of reactor for the treatment of domestic sewage.

The anaerobic process through UASB reactors presents several advantages in relation to conventional aerobic processes, especially when applied in warm-climate locations, such as most of the developing countries. In these situations, a system can have the following main characteristics:

- compact system, with low land requirements
- low construction and operating costs
- low sludge production
- low energy consumption (just for the influent pumping station, when necessary)
- satisfactory COD and BOD removal efficiencies, amounting to 65 to 75%
- high concentration and good dewatering characteristics of the excess sludge

Although the UASB reactors present many advantages, there are still some disadvantages or limitations:

- possibility of release of bad odours
- low capacity of the system in tolerating toxic loads
- long time interval necessary for the start-up of the system
- need for a post-treatment stage

In situations in which the wastewater is predominantly domestic, the presence of sulfur compounds and toxic materials usually occurs at very low levels, being well handled by the treatment system. When well designed, constructed and operated, the system should not present bad smell and failure problems due to the presence of toxic elements and/or inhibitors.

The start-up of the system can be slow (4 to 6 months), but only in situations in which seed sludge is not used. In the past few years, with the use of well-based start-up methodologies and the establishment of appropriate operational routines, significant progresses were achieved towards reducing the start-up period of the systems and minimising the operational problems in this phase. In situations already reported (Chernicharo and Borges, 1996), in which small amounts of seed sludge were used (less than 4% of the reactor volume), the start-up period was reduced to 2 or 3 weeks. In any case, the quality of the biomass to be developed in the system will depend on an appropriate operational routine and, consequently, on the stability and efficiency of the treatment process.

However, apart from the great advantages of the UASB reactors, the quality of the effluent produced usually does not comply with most discharge standards established by environmental agencies. Until recent years there were not many experiences that consolidate an overall view of the combined stages of *anaerobic treatment* and *post-treatment*. However, important advances have been achieved recently, as mentioned by Chernicharo *et al.* (2001b).

The design of UASB reactors is very simple and does not require the installation of any sophisticated equipment or packing medium for biomass retention. In spite of the accumulated knowledge on UASB reactors, there are still no clear, systematised guidelines accessible by designers for the design of these reactors. It is important that the several design criteria and parameters for UASB reactors are expressed in a clear and sequential manner, allowing the dimensioning of the reaction, sedimentation and gas capture chambers.

5.2.2 Process principles

The reactor is initially inoculated with sufficient quantities of anaerobic sludge, and its low-rate feeding is started soon afterwards, in the upflow mode. This initial period is referred to as *start-up* of the system, being the most important phase of the operation of the reactor. The feeding rate of the reactor should be increased progressively, according to the success of the system response. After some months of operation, a highly concentrated *sludge bed* (4 to 10%, that is, 40 to 100 gTS/L)

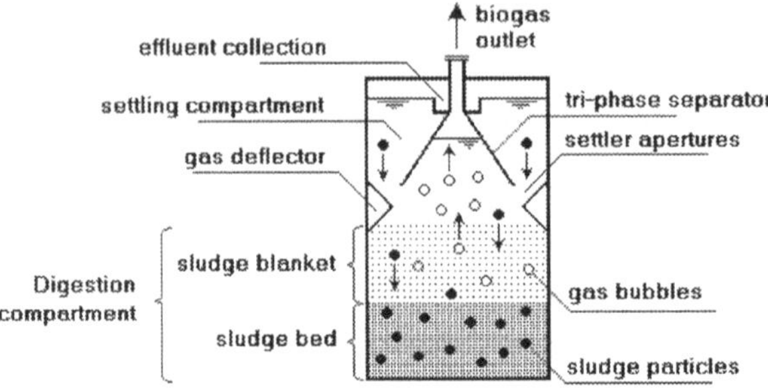

Figure 5.7.   Schematic drawing of a UASB reactor

is developed close to the bottom of the reactor. The sludge is very dense and has excellent settling characteristics. The development of sludge granules (diameters from 1 to 5 mm) may occur, depending on the nature of the seeding sludge, on the characteristics of the wastewater and on the operational conditions of the reactor.

An area of more dispersed bacterial growth, named *sludge blanket*, is developed above the sludge bed, with solids presenting lower concentrations and settling velocities. The concentration of sludge in this area usually ranges between 1 and 3%. The system is self-mixed by the upflow movement of biogas bubbles and by the liquid flow through the reactor. During the start-up of the system, when the biogas production is usually low, some form of additional mixing, such as by the recirculation of gas or effluent, may become necessary. Substrate is removed throughout the bed and sludge blanket, although removal is more pronounced at the sludge bed.

The sludge is carried by the upflow movement of the gas bubbles, and the installation of a *three-phase separator* (gases, solids and liquids) in the upper part of the reactor is necessary, to allow sludge retention and return. There is a sedimentation chamber around and above the three-phase separator, where the heaviest sludge is removed from the liquid mass and returned to the digestion compartment, while the lightest particles leave the system together with the final effluent (see Figure 5.7).

The installation of the gas, solids and liquid separator guarantees the return of the sludge and the high retention capacity of large amounts of high-activity biomass, with no need for any type of packing medium. As a result, UASB reactors present high solids residence times (sludge age), much higher than the hydraulic detention times, which is a characteristic of the high-rate anaerobic systems. Sludge ages in UASB reactors usually exceed 30 days, leading to stabilisation of the excess sludge removed from the system.

The UASB reactor is capable of supporting high organic loading rates and the great difference, when compared with other reactors of the same generation, is its constructive simplicity and low operational costs. The most important principles that govern the operation of UASB reactors are:

- the upward flow should assure a maximum contact between the biomass and the substrate
- short circuits should be avoided, to allow retention times sufficient for the degradation of the organic matter
- the system should have a well designed device capable of separating suitably the biogas, the liquid and the solids, releasing the first two and allowing the retention of the last
- the sludge should be well adapted, with high specific methanogenic activity and excellent settling characteristics. If possible, the sludge should be granulated, once this type of sludge presents much better characteristics than those of the flocculent sludge

5.2.3 Typical configurations

UASB reactors were initially designed for the treatment of industrial effluents as cylindrical or prismatic-rectangular structures, where the areas of the digestion and sedimentation compartments were equal, therefore forming vertical wall reactors. The adaptation of these reactors to the treatment of low-concentration wastewater (such as domestic sewage) has led to different configurations, in view of the following main aspects:

- In the design of UASB-type reactors treating low-concentration sewage, the design is ruled by the hydraulic loading criteria, and not by the organic loading criteria, as discussed in the following item. In this situation, the upward velocity in the digestion and sedimentation compartments becomes essentially important: excessive velocities result in the loss of biomass from the system, thus reducing the stability of the process. Consequently, the height of the reactor should be reduced and its cross section should be increased, to keep the upward velocities within suitable ranges (see Table 5.14).
- For reactors treating industrial effluents, the influent is usually distributed from the bottom of the reactor, unlike reactors treating domestic sewage, where the influent distribution device is located in the upper part of the reactor (see Figures 5.8 to 5.10). Consequently, the surface area of the sedimentation compartment may be reduced in view of the area occupied by the influent distribution device. Thus, depending on the hydraulic loads applied to the system, it may be necessary to use larger cross sections close to the sedimentation compartment, to reduce the upward velocities and enable the sedimentation of the sludge in this compartment. In this case, the reactor adopts a variable section, smaller close to the digestion compartment and larger close to the sedimentation compartment (see Figure 5.9).
- The implementation of an equalisation tank is usually planned upstream the UASB reactor in the treatment of industrial effluents, allowing its operation to be carried out within more uniform flow and organic loading ranges.

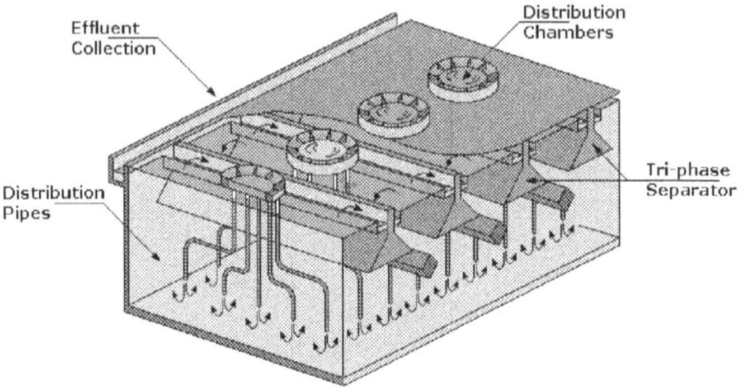

Figure 5.8.　Schematic representation of a rectangular UASB reactor

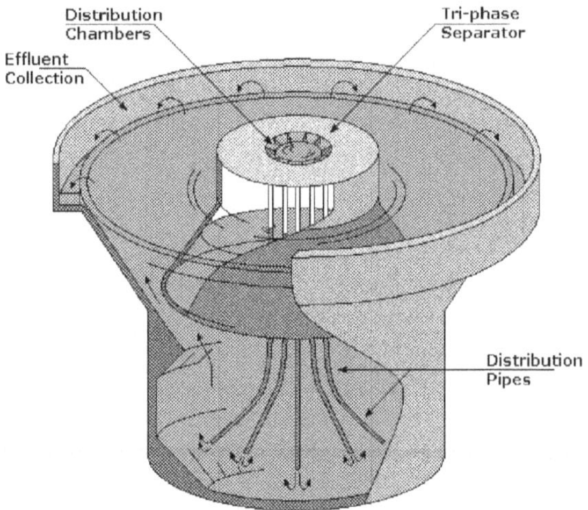

Figure 5.9.　Schematic representation of a circular UASB reactor

On the other hand, the influent to a domestic sewage treatment plant undergoes no equalisation (unless there is a pumping station), exposing the UASB reactor to flow and load variations that may be extremely high. Once again, the increased cross section of the reactor close to the sedimentation compartment may be a necessary strategy to guarantee low upward velocities during peak flows.

The shape of the reactors in plan can be either circular or rectangular. Circular reactors are more economical from the structural point of view, being used more for small populations, usually with a single unit. Rectangular reactors are more suitable for larger populations, when modulation becomes necessary, once

Figure 5.10.   View of a full-scale UASB reactor
Source: Ipatinga WWTP, COPASA, Brazil

a wall can serve two contiguous modules. Figures 5.8 and 5.9 illustrate two typical configurations of UASB reactors, a rectangular one and a circular one. Figure 5.10 shows a full-scale rectangular UASB reactor.

5.2.4 Design criteria

One of the most important aspects of the anaerobic process applying UASB reactors is its ability to develop and maintain high-activity sludge of excellent settling characteristics. For this purpose, several measures should be taken in relation to the design and operation of the system.

The main design criteria for reactors treating organic wastes of either domestic or industrial nature are presented below. Specific criteria should be adopted for certain types of industrial effluents in view of the concentration of the influent wastewater, the presence of toxic substances, the amount of inert and biodegradable solids and other aspects.

(a)   Volumetric hydraulic load and hydraulic detention time

The volumetric hydraulic load is the amount (volume) of wastewater applied daily to the reactor, per unit of volume. The hydraulic detention time is the reciprocal of the volumetric hydraulic load,

$$VHL = \frac{Q}{V} \tag{5.8}$$

where:
$\quad$ VHL = volumetric hydraulic load ($m^3/m^3 \cdot d$)
$\quad\quad$ Q = flowrate (m^3/d)
$\quad\quad$ V = total volume of the reactor (m^3)

$$t = \frac{1}{VHL} \qquad (5.9)$$

where:

   t = hydraulic detention time (d)

or

$$\boxed{t = \frac{V}{Q}} \qquad (5.10)$$

Experimental studies demonstrated that the volumetric hydraulic load should not exceed the value of 5.0 $m^3/m^3{\cdot}d$, which is equivalent to a minimum hydraulic detention time of 4.8 hours.

The design of reactors with higher hydraulic loading values (or lower hydraulic detention times) can be detrimental to the operation of the system in relation to the following main aspects:

- excessive loss of biomass, that is washed out with the effluent, due to the resulting high upflow velocities in the digestion and settling compartments
- reduced solids retention time (sludge age), and a consequently decreased degree of stabilisation of the solids
- possibility of failure in the system, once the biomass residence time in the system becomes shorter than its growth rate

As shown previously, the hydraulic detention time parameter (t) is of fundamental importance, since it is directly related to the velocity of the anaerobic digestion process, and that, in turn, depends on the size of the reactor. For average temperatures close to 20 °C, the hydraulic detention time can vary from 6 to 16 hours, depending on the type of wastewater. Pilot-scale studies with reactors operated at an average temperature of 25 °C and fed with domestic sewage with relatively high alkalinity showed that a 4-hour hydraulic detention time did not affect the performance of these reactors or their operational stability (van Haandel and Catunda, 1998).

Hydraulic detention times ranging from **8 to 10 hours**, considering the daily average flowrate, have been adopted for the treatment of *domestic sewage* at a temperature of approximately 20 °C. The detention time for the maximum flowrate should not be shorter than 4 hours, and the maximum flow peaks should not be longer than 4 to 6 hours. Table 5.3 presents some guidelines for the establishment of hydraulic detention times in designs of UASB reactors treating domestic sewage.

Thus, knowing the influent flowrate and assuming a certain design hydraulic detention time, the volume of the reactor can be calculated by Equation 5.10, rearranged as follows:

$$\boxed{V = Q.t} \qquad (5.11)$$

Table 5.3. Recommended hydraulic detention times for UASB reactors treating domestic sewage

Sewage temperature (°C)	Hydraulic detention time (hour)	
	Daily average	Minimum (during 4 to 6 hour)
16 to 19	>10 to 14	>7 to 9
20 to 26	>6 to 9	>4 to 6
>26	>6	>4

Source: Adapted from Lettinga and Hulshoff Pol (1991)

(b) Organic loading rate

The volumetric organic load is defined as the amount (mass) of organic matter applied daily to the reactor, per volume unit:

$$L_v = \frac{Q \times S_0}{V} \tag{5.12}$$

where:

L_v = volumetric organic loading rate (kgCOD/m^3·d)
Q = flowrate (m^3/d)
S_0 = influent substrate concentration (kgCOD/m^3)
V = total volume of the reactor (m^3)

Hence, knowing the flowrate and the concentration of the influent wastewater, and assuming a certain design volumetric organic load (L_v), the volume of the reactor can be calculated by Equation 5.12, rearranged as follows:

$$\boxed{V = \frac{Q \times S_0}{L_v}} \tag{5.13}$$

In the case of *industrial effluents* with a high concentration of organic matter, literature reports extremely high organic loads successfully applied to pilot facilities (45 kgCOD/m^3·d), although the organic loads adopted in the design of full-scale plants have been, as a rule, **lower than 15 kgCOD/m^3·d**. For such effluents, the volumetric organic load to be applied is what defines the reactor volume. Concerning *domestic sewage* with a relatively low concentration of organic matter (usually lower than 1,000 mgCOD/L), the volumetric organic load to be applied is much lower, ranging from **2.5 to 3.5 kg COD/m^3·d**; higher values result in excessive hydraulic loads and, consequently, in excessive upflow velocities. In this case, as stated previously, the reactor should be designed considering the volumetric hydraulic load. For example, Figure 5.11 illustrates the relation between wastewater concentration and the criteria used to determine the volume of the reactor, considering the following established data: t = 8 hours, L_v = 15 kgCOD/m^3·d and Q = 250 m^3/hour.

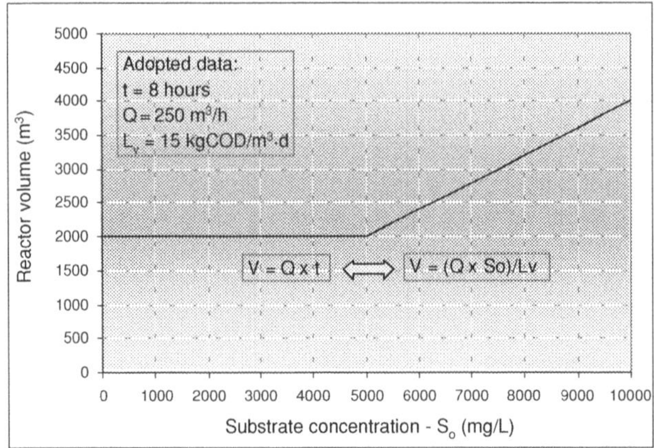

Figure 5.11.   Relation between wastewater concentration and reactor volume (adapted from Lettinga and Hulshoff Pol, 1995)

(c)   Biological loading rate (sludge loading rate)

The biological or sludge loading rate refers to the amount (mass) of organic matter applied daily to the reactor, per unit of biomass present:

$$L_s = \frac{Q \times S_0}{M} \qquad (5.14)$$

where:

L_s = biological or sludge loading rate (kgCOD/kgVS·d)
Q = average influent flowrate (m³/d)
S_0 = influent substrate concentration (kgCOD/m³)
M = mass of microorganisms present in the reactor (kgVS/m³)

The procedures to determine the amount of biomass in the reactor were covered in Chapter 3.

Literature recommends that the initial biological loading rate during the start-up of an anaerobic reactor should range from 0.05 to 0.15 kgCOD/kgVS·d, depending on the type of effluent being treated. These loads should be gradually increased, according to the efficiency of the system.

The maximum biological loading rate depends on the methanogenic activity of the sludge. For domestic sewage, the methanogenic activity usually ranges from 0.3 to 0.4 kgCOD/kgVS·d, which is, therefore, the limit for the biological load.

Recent experiments with UASB reactors treating domestic sewage indicated that the application of biological loading rates ranging from 0.30 to 0.50 kgCOD/kgVS·d during the start-up of the system did not harm the stability of the process in terms of pH and volatile fatty acids.

(d)   Upflow velocity and reactor height

The upflow velocity of the liquid is calculated from the relation between the influent flowrate and the cross section of the reactor, as follows:

$$\boxed{v = \frac{Q}{A}}$$

(5.15)

where:

v = upflow velocity (m/hour)

Q = flow (m^3/hour)

A = area of the cross section of the reactor, in this case the surface area (m^2)

or alternatively, from the ratio between the height and the HDT:

$$v = \frac{Q \times H}{V} = \frac{H}{t}$$

(5.16)

where:

H = height of the reactor (m)

The maximum upflow velocity in the reactor depends on the type of sludge present and on the loads applied. For reactors operating with flocculent sludge and organic loading rates ranging from 5.0 to 6.0 kgCOD/m^3·d, the average upflow velocities should amount to **0.5 to 0.7 m/hour**, with temporary peaks up to 1.5 to 2.0 m/hour being tolerated for 2 to 4 hours. For reactors operating with granular sludge, the upflow velocities can be significantly higher, amounting to 10 m/hour. For the treatment of domestic sewage, the upflow velocities presented in Table 5.4 are recommended.

A close relation between the upflow velocity the height of the reactor and the hydraulic detention time can be verified in Equation 5.16, as shown in Figure 5.12. For the upflow velocities (v) and the hydraulic detention times (t) recommended for the design of UASB reactors treating domestic sewage (v usually lower than 1.0 m/hour for Q_{av} and t between 6 and 10 hours for temperatures ranging between 20 and 26 °C), reactor depths should be between **3 and 6 m**.

Table 5.4. Upflow velocities recommended for the design of UASB reactors treating domestic sewage

Influent flowrate	Upflow velocity (m/hour)
Average flow	0.5 to 0.7
Maximum flow	<0.9 to 1.1
Temporary peak flows [*]	<1.5

[*] flowrate peaks lasting 2 to 4 hours
Source: Adapted from Lettinga and Hulshoff Pol (1995)

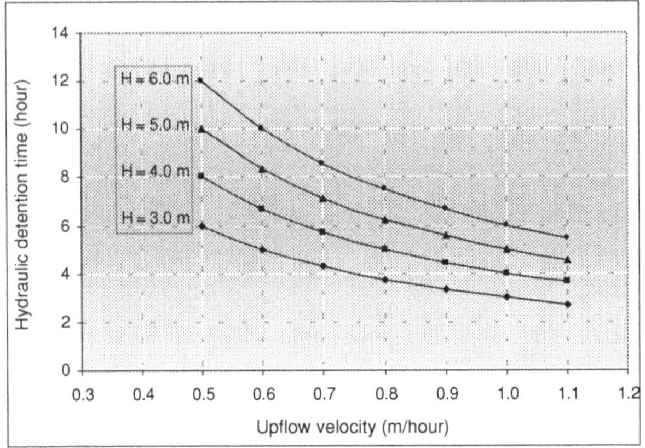

Figure 5.12.   Relation between upflow velocity and HDT for different reactor heights

(e)   UASB reactor efficiencies

Mathematical models applied to the design and operation of anaerobic systems have still been little used in practice, particularly for systems treating complex substrates such as domestic sewage, although valuable achievements in this field are expected in the following years with the release of the Anaerobic Digestion Model No. 1 (Batstone *et al.*, 2002), developed by the IWA task group for mathematical modelling of anaerobic digestion processes. Meanwhile, the efficiency of UASB reactors is estimated mainly by means of empirical relations, obtained from experimental results of systems in operation.

Figures 5.13 and 5.14 show the operational results of 16 full-scale UASB reactors, all of them operating within the temperature range between 20 and 27 °C, influent COD between 300 and 1,400 mg/L and influent BOD between 150 and 850 mg/L. It can be noted that the COD and BOD removal efficiencies are

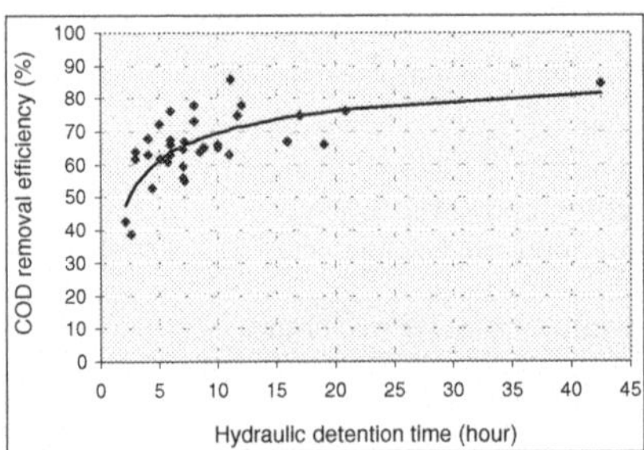

Figure 5.13.   COD removal efficiencies in UASB reactors treating domestic sewage

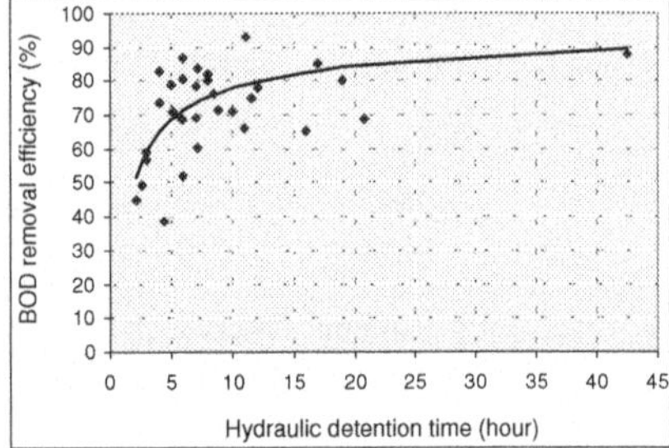

Figure 5.14.   BOD removal efficiencies in UASB reactors treating domestic sewage

substantially affected by the hydraulic detention time of the system, ranging from 40 to 70% for COD removal and from 45 to 90% for BOD removal.

From the fitting of the operational results of these 16 reactors, efficiency curves were obtained and represented by Equations 5.17 and 5.18. These equations make it possible to estimate the COD and BOD removal efficiencies of UASB reactors treating domestic sewage under tropical conditions (wastewater temperature within 20 and 27 °C) as a function of the hydraulic detention time. However, their limitation should be emphasised due to the small number of data used for the determination of the empirical constants, which showed great deviations amongst themselves.

$$E_{COD} = 100 \times (1 - 0.68 \times t^{-0.35}) \tag{5.17}$$

where:

E_{COD} = efficiency of the UASB reactor in terms of COD removal (%)
   t = hydraulic detention time (hour)
0.68 = empirical constant
0.35 = empirical constant

$$E_{BOD} = 100 \times (1 - 0.70 \times t^{-0.50}) \tag{5.18}$$

where:

E_{BOD} = efficiency of the UASB reactor in terms of BOD removal (%)
   t = hydraulic detention time (hour)
0.70 = empirical constant
0.50 = empirical constant

Estimation of the COD and BOD concentrations in the final effluent

From the efficiency expected for the system, the COD and BOD concentration in the final effluent can be estimated as follows:

$$C_{effl} = S_0 - \frac{E \times S_0}{100} \tag{5.19}$$

where:

C_{effl} = effluent total COD or BOD concentration (mg/L)
  S_0 = influent total COD or BOD concentration (mg/L)
  E = COD or BOD removal efficiency (%)

Estimation of the SS concentration in the final effluent

The concentration of suspended solids in the final effluent from UASB reactors depends on a series of factors, including:

- the concentration and the settling characteristics of the sludge present in the reactor
- the sludge wastage frequency and the height of the sludge bed and blanket in the reactor

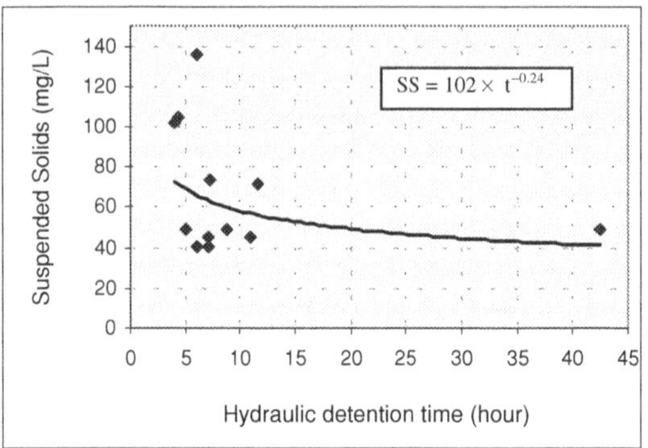

Figure 5.15.   SS concentrations in the effluent from UASB reactors treating domestic sewage

- the velocities through the apertures to the sedimentation compartment
- the presence of scum baffles in the sedimentation compartment
- the efficiency of the gas, solids and liquid separator
- the loading rates and the hydraulic detention times in the digestion and sedimentation compartments

In the absence of studies that relate, in a systematised manner, the concentration of solids in the effluent to some of the factors previously mentioned, option was made for the consolidation of the operational results of five reactors taking into account only the hydraulic detention time in the system (see Figure 5.15). The results of solids from the other 11 reactors, which were analysed for COD and BOD removal efficiencies, were not included because they were unusual or not available. It can be observed that the effluent solids concentrations, which varied from 40 to 140 mg/L, were affected by the hydraulic detention time within the system.

From the fitting of the operational results of the five reactors, a curve representing the expected concentration of solids in the effluent was obtained (Equation 5.20). Likewise for COD and BOD, the limitation of this expression is emphasised due to the very reduced number of data used to determine the empirical constants and also to the great deviations existing amongst the data. Besides that, other variables that interfere with the concentration of solids in the effluent are not considered in Equation 5.20.

$$SS = 102 \times t^{-0.24} \tag{5.20}$$

where:

SS = effluent suspended solids concentration (mg/L)
t = hydraulic detention time (hour)
102 = empirical constant
0.24 = empirical constant

(f)   Influent distribution system

To obtain a good performance from UASB reactors, it is essential that the influent substrate is evenly distributed in the lower part of the reactors, to ensure a close contact between the biomass and the substrate. For that purpose and so that the maximum advantage is taken from the biomass present in the reactors, it is essential that preferential pathways (hydraulic short circuits) are avoided through the sludge bed as much as possible. That is particularly important when the process is used in the treatment of low-concentration (such as domestic sewage) and/or low-temperature sewage, once in those situations the biogas production can be very low to allow appropriate mixing within the digestion compartment. Other potential risks for the occurrence of short circuits are:

- short height of the sludge bed
- small number of influent distributors
- occurrence of very concentrated sludge with very high settling velocities

Distribution compartments

An even distribution of the influent is very important in UASB reactors, to ensure a better mixing regime and a reduced occurrence of dead zones on the sludge bed. Thus, the equal division of the influent flow to the several distributing tubes should be done by small compartments (boxes) fed by weirs. Each box feeds a single distribution tube extending to the bottom of the reactor. These compartments, installed in the upper part of the reactor, ensure the uniform distribution of sewage throughout the bottom of the tank, besides enabling the visualisation of occasional increments in the head loss, in each distributor. Once an increased head loss is detected in a distributor, the tube can be easily unblocked by using appropriate rods. Examples of influent distribution structures in UASB reactors are presented in Figures 5.16 and 5.17.

Distribution tubes

Wastewater is routed from the distribution compartments to the bottom of the reactor through distribution tubes. The main requirements for these tubes are as follows:

- the diameter should be large enough to enable a descending sewage velocity **lower than 0.2 m/s**, so that the air bubbles occasionally dragged to inside the tube can go back upwards (opposite the direction of the sewage). The introduction of air bubbles in the reactor should be avoided for the following reasons (van Haandel and Lettinga, 1994): (i) they may cause the aeration of the anaerobic sludge, harming methanogenesis; and (ii) they may cause a potentially explosive mixture with the biogas accumulated close to the three-phase separator. In the case of treatment of low-concentration sewage, this velocity requirement is usually met when the tubes have a 75 mm diameter.

Figure 5.16.   Influent distribution structure in a circular reactor (source: Nova Vista WWTP, SAAE Itabira, Minas Gerais, Brazil)

Figure 5.17.   Influent distribution structure in a rectangular reactor (source: Ipatinga WWTP, COPASA, Brazil)

- the diameter should be large enough to prevent the solids present in the influent from frequently obstructing the tubes. In this aspect, the excessive presence of solids in the influent can increase the obstruction frequency of the distribution tubes, and the planning of an efficient screening system for the previous removal of solids is essential. Practical experience has shown that distribution tubes with diameters of **75 and 100 mm** meet this requirement.

- the diameter should be small enough to allow a higher flow velocity at its lower end (bottom of the reactor), which favours good mixing and greater contact with the sludge bed. Besides that, a higher velocity helps avoid the deposition of inert solids close to the discharge point of the tube. This requirement is somehow incompatible with the previous ones, once a reduced diameter of the tube hinders the upward movement and the release of air bubbles, besides increasing their possibilities of blocking. A solution that can be adopted is the reduction of the tubing section just close to its lower end, thus keeping an area large enough to avoid blockage. In the case of treatment of domestic sewage, practical experience has shown that *nozzles* with a diameter of approximately **40 to 50 mm** can be used with the purpose of increasing the velocity in the piping exit. For these diameters, the exit velocities are usually higher than 0.40 m/s, which is enough to avoid the deposition of sand close to the ends of the tubes. As an alternative to the nozzles, apertures (windows) can be made on the side

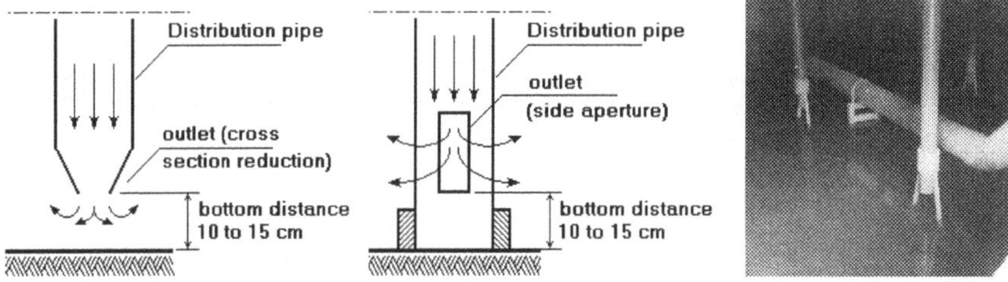

Figure 5.18.   Examples of distribution tube ends

ends of the distribution tubes. In this case, two openings with a **25 mm** × **40 mm** cross section can be used, creating an area corresponding to a 50 mm nozzle. These devices are illustrated in Figure 5.18.

The lower ends of the distribution tubes should be installed at pre-established points, according to the influence area defined in the design. The maintenance of a fixed position in relation to the bottom of the reactor is important.

Number of distribution tubes

As previously mentioned, the correct distribution of the incoming sewage is one of the most important aspects for the correct operation of the reactor, to ensure an effective contact with the biomass present in the reactor. The number of distribution tubes is determined according to the area of the cross section of the reactor and the influence area adopted for each distributor, as follows:

$$N_d = \frac{A}{A_d} \tag{5.21}$$

where:
N_d = number of distribution tubes
A = area of the cross section of the reactor (m^2)
A_d = influence area of each distributor (m^2)

Preliminary guidelines are presented in Table 5.5 for the influence area of flow distributors in UASB reactors, as a function of the type of sludge and organic loads applied to the system.

In the case of reactors treating domestic sewage, a flocculent sludge is usually developed in the system, with medium to high concentration characteristics. The organic loads applied to the system generally amount from 1.0 to 3.0 kgCOD/m^3·d. In these situations, and according to the guidelines presented in Table 5.5, the influence area of each distributor should be from **1.5 to 3.0 m^2**.

According to a survey done by van Haandel and Lettinga (1994), influence areas of distributors ranging from 1.0 to 4.0 m^2 have been used, as presented in Table 5.6.

Table 5.5. Preliminary guidelines for the influence area of flow distributors in UASB reactors

Sludge type	Organic load applied $(kgCOD/m^3 \cdot d)$	Influence area of each distributor (m^2)
Dense and flocculent	<1.0	0.5 to 1.0
(concentration >40 kgTSS/m^3)	1.0 to 2.0	1.0 to 2.0
	>2.0	2.0 to 3.0
Relatively dense and flocculent	<1.0 to 2.0	1.0 to 2.0
(concentration 20 to 40 kgTSS/m^3)	>3.0	2.0 to 5.0
	<2.0	0.5 to 1.0
Granular	2.0 to 4.0	0.5 to 2.0
	>4.0	>2.0

Source: Lettinga and Hulshoff Pol (1995)

Table 5.6. Influence areas of flow distributors in UASB reactors treating domestic sewage

System	Influence area of each distributor (m^2)
Itabira (Minas Gerais, Brazil)	2.3 to 3.0
Pedregal (Paraíba, Brazil)	2.0 to 4.0
São Paulo (Cetesb, Brazil)	2.0
Bucaramanga (Colombia)	2.9
Cali (Colombia)	1.0 to 4.0
Kampur (India)	3.7

Source: Adapted from van Haandel and Lettinga (1994)

However, there have been designs that consider an influence area larger than 4 to 5 m^2 for each distribution tube. In these cases, the mixing regime can be affected during the operation of the reactor, harming the contact between biomass and substrate and favouring the creation of dead zones on the sludge bed. Consequently, the efficiency expected for the process may not be reached.

In the particular case of trunk-conical reactors, the influence area of the distribution tubes is not uniform over the height of the digestion compartment, once the cross section of the reactor increases with its height. In these cases, the calculations should consider the cross section close to the deepest part of the reactor (where the sludge bed, more concentrated, is located), that is, close to the first metre of depth of the reactor, to ensure an influence area suitable for the flow distributors.

Considering the low cost of the distribution tubes and the substantial benefits resulting from a correct distribution system, it is recommended that the influence areas of each distributor range from **2.0 to 3.0 m^2** for the treatment of domestic sewage with typical COD concentrations (400 to 600 mg/L).

(g)   Three-phase separator

The gas, solids and liquid separator (three-phase separator) is an essential device that needs to be installed in the upper part of the reactor. The main objective of this separator is to maintain the anaerobic sludge inside the reactor, allowing the system to be operated with high solids retention times (high sludge age). This is initially achieved by separating the gas contained in the liquid mixture, enabling, as a consequence, the maintenance of optimal settling conditions in the settling compartment. Once the gas is effectively removed, the sludge can be separated from the liquid in the settling compartment, and then returned to the digestion compartment.

Separation of gases

The design of the gas, solids and liquid separating device (three-phase separator) depends on the characteristics of the wastewater, the type of sludge present in the reactor, the organic load applied, the expected biogas production and the dimensions of the reactor. Aiming at avoiding sludge flotation and the consequent biomass loss from the reactor, the dimensions of the separator should be such that they allow the formation of a liquid–gas interface inside the gas collector sufficient to allow the easy release of the gas entrapped in the sludge. The biogas release rate should be high enough to overcome a possible scum layer, but low enough to quickly release the gas from the sludge, not allowing the sludge to be dragged and consequently accumulated in the gas exit piping. Souza (1986) recommends minimum release rates of 1.0 $m^3gas/m^2 \cdot$hour and maximum rates from 3.0 to 5.0 $m^3gas/ m^2 \cdot$hour. The biogas release rate is established by the following equation:

$$K_g = \frac{Q_g}{A_i} \tag{5.22}$$

where:

$\quad K_g$ = biogas release rate ($m^3/m^2 \cdot$hour)
$\quad Q_g$ = expected biogas production ($m^3/$hour)
$\quad A_i$ = area of the liquid–gas interface (m^2)

Evaluation of the biogas production

The biogas production can be evaluated from the estimated influent COD load to the reactor that is converted into methane gas, according to Chapter 2. In a simplified manner, the portion of COD converted into methane gas can be determined as follows:

$$COD_{CH_4} = Q \times (S_0 - S) - Y_{obs} \times Q \times S_0 \tag{5.23}$$

where:

$\quad COD_{CH_4}$ = COD load converted into methane ($kgCOD_{CH_4}/d$)
$\quad Q$ = average influent flow (m^3/d)
$\quad S_0$ = influent COD concentration ($kgCOD/m^3$)
$\quad S$ = effluent COD concentration ($kgCOD/m^3$)
$\quad Y_{obs}$ = coefficient of solids production in the system, in terms of COD (0.11 to 0.23 $kgCOD_{sludge}/kgCOD_{appl}$).

The methane *mass* ($kgCOD_{CH_4}/d$) can be converted into *volumetric* production ($m^3 CH_4/d$) by using the following equations:

$$Q_{CH_4} = \frac{COD_{CH_4}}{K(t)} \qquad (5.24)$$

where:

Q_{CH_4} = volumetric methane production (m^3/d)
$K(t)$ = correction factor for the operational temperature of the reactor ($kgCOD/m^3$)

$$K(t) = \frac{P \times K_{COD}}{R \times (273 + T)} \qquad (5.25)$$

where:

P = atmospheric pressure (1 atm)
K_{COD} = COD corresponding to one mole of CH_4 (64 gCOD/moL)
R = gas constant (0.08206 atm·L/mole·K)
T = operational temperature of the reactor (°C)

Once the theoretical methane production is obtained, the total biogas production can be estimated from the expected methane content. For the treatment of domestic sewage, the methane fraction in the biogas usually ranges from 70 to 80%.

Separation of solids

After the separation of the gases, the liquid and the solid particles that leave the sludge blanket have access to the sedimentation compartment. Ideal conditions for sedimentation of the solid particles occur in this compartment, due to the low upflow velocities and the absence of gas bubbles. The return of the sludge retained in the sedimentation compartment to the digestion compartment does not require any special measure, as long as the following basic guidelines are met:

- installation of deflectors, located immediately below the apertures to the sedimentation compartment, to enable the separation of the biogas, and allow only liquid and solids to enter the sedimentation compartment
- construction of the sedimentation compartment walls with slopes always higher than 45°. Ideally, slopes equal to or higher than 50° should be adopted
- adoption of depths of the sedimentation compartment ranging from 1.5 to 2.0 m
- adoption of surface loading rates and hydraulic detention times in the sedimentation compartment according to Table 5.7

Table 5.7. Surface loading rates and hydraulic detention times in the sedimentation compartment

Influent flow	Surface loading rate (m/hour)	Hydraulic detention time (hour)
Average flow	0.6 to 0.8	1.5 to 2.0
Maximum flow	<1.2	>1.0
Temporary peak flows[*]	<1.6	>0.6

(*) Peak flow lasting between 2 and 4 hours

Apertures to the sedimentation compartment

The apertures that allow the passage of wastewater to the sedimentation compartment should be designed to allow:

- the separation of the gases before the sewage has access to the sedimentation zone, favouring the sedimentation of the solids in the settler compartment. For that purpose, the design of the apertures should allow an appropriate overlap of the gas deflector, to ensure the correct separation of the gas and liquid phases
- the retention of solids in the digestion compartment, by maintaining velocities in the apertures lower than those recommended in Table 5.8
- the return of the solids retained in the sedimentation compartment to the digestion compartment. This return should occur when appropriate slopes of the walls of the sedimentation compartment and gas deflectors are adopted, and also by maintaining compatible velocities through the apertures

Hydraulic detention time in the sedimentation compartment

The hydraulic detention time recommended in the sedimentation compartment ranges from 1 to 2 hours, as presented in Table 5.7. Verifications made in projects already implemented have indicated that the detention times for average flows are not always within the established range. For reactors fed by pumping stations, the detention times tend to be even more reduced, sometimes reaching 0.5 hour when there are two or more pumps in operation.

Table 5.8. Velocities in the apertures to the sedimentation compartment

Influent flow	Velocity (m/hour)
Average flow	<2.0 to 2.3
Maximum flow	<4.0 to 4.2
Temporary peak flows[*]	<5.5 to 6.0

(*) Peak flows lasting between 2 and 4 hours

Figure 5.19.   Effluent collection device (plate with V-notch weirs)
Source: Nova Vista WWTP, Itabira, Minas Gerais, Brazil

In situations in which the velocities through the apertures are high and the detention time in the sedimentation compartment is small, a high loss of solids in the effluent and the eventual failure of the treatment system are expected.

(h)   Effluent collection

The effluent is collected from the reactor in its upper part, within the sedimentation compartment. The devices usually used for the collection of effluent are plates with V-notch weirs and submerged perforated tubes.

If a launder with V-notch weirs is used (see Figure 5.19), special care should be taken with their levelling, once small slopes in the launder can represent a significant variation in the flow collected at different points. A scum baffle, submerged at approximately 20 cm, should be included along the launder. Additional care regarding the launders refers to the possibility of gas release, particularly H_2S, in view of the turbulence close to the weirs. In this sense, submerged outlets, with no effluent turbulence, are more suitable.

The alternative of using submerged perforated tubes for the collection of effluent has been shown to be very efficient, mainly in three aspects:

- as they are submerged devices, the maintenance of uniform flows in the holes is favoured, and the levelling requirements of the collecting tubes are less important
- the use of submerged tubes decreases or eliminates the risks of turbulence, as well as of release of gases and bad odours
- the submerged collection does not require the use of scum baffles, once the effluent is removed below the scum layer

One of the disadvantages of the collection system by submerged tubes is the possible accumulation of solids in the holes and inside the piping. As cleaning is not always possible, it is recommended that the tubes are laid with a minimum slope of 1%, so that they may be self-cleaned.

(i) Gas system

The uncontrolled release of biogas into the atmosphere is detrimental, not just for the possible occurrence of bad odours in the neighbourhood, but mainly for the risks inherent to the methane gas, which is combustible. Thus, the biogas produced in the reactor should be collected, measured and, later, either used or burnt. The biogas removal system from the liquid–gas interface inside the reactor consists of:

- collecting piping
- sealed compartment with hydraulic seal and biogas purge
- biogas meter
- biogas reservoir

When the biogas is not used, the gas reservoir is replaced by a security valve and a gas burner, preferably located at a safe distance from the reactor, as illustrated in Figures 5.20 and 5.21.

The biogas flow meter is an important device for the monitoring of the amount of gas produced in the system, being essential for the evaluation of the process efficiency. To avoid damage to the meters, caused by the dragging of condensed liquids into the collecting piping, the average biogas flow velocity should not exceed 3.5 m/s.

Further considerations on the collection, treatment and destination of the gases generated in anaerobic reactors can be found in Campos and Pagliuso (1999), Belli Filho *et al.* (2001), Andreoli *et al.* (2003) and Cassini *et al.* (2003).

(j) Sludge sampling and discharge system

The design of the reactor should comprise a group of valves and piping that allows both sampling and discharge of the solids present in the reactor.

Sludge sampling system

The sampling system usually consists of a series of valves installed along the height of the digestion compartment, to enable the monitoring of the growth and quality

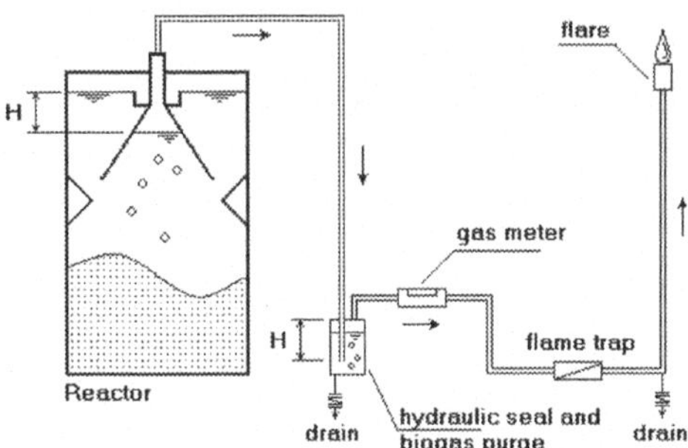

Figure 5.20.   Diagram of a gas system in UASB reactors

Figure 5.21.   View of a hydric seal and gas burner (source: Ipatinga WWTP, Minas Gerais, Brazil)

of the biomass in the reactor. One of the most important operational routines in the treatment system is the evaluation of the amount and activity of the biomass present in the reactor, by means of two basic mechanisms:

- determination of the solids profile and mass of microorganisms present in the system, as exemplified in Chapter 3 (Example 3.1)
- evaluation of the specific methanogenic activity of the biomass, as exemplified in Chapter 3 (Example 3.2)

The continuous monitoring of the biomass present in the reactor will allow the operation personnel to have more control actions over the solids in the system, such as:

- identification of the height and concentration of the sludge bed in the reactor, allowing the establishment of discharge strategies (discharge amount and frequency)
- determination of the ideal sludge discharge points, according to the results of the specific methanogenic activity tests and the characteristics of the sludge

Thus, to enable the removal and characterisation of the biomass at different levels of the digestion compartment, the installation of valves is recommended, from the base of the reactor, with the following characteristics:

- spacing: 50 cm
- diameter: $1\frac{1}{2}$ to 2 inches
- type: ball valve

Sludge withdrawal system

The sludge discharge system is intended for the periodical removal of the excess sludge produced in the reactor, also allowing the removal of inert material that may accumulate at the bottom of the reactor. At least two sludge withdrawal points should be planned, one close to the bottom of the reactor and another approximately 1.0 to 1.5 m above the bottom (depending on the height of the digestion compartment), to allow a higher operational flexibility. A minimum diameter of 100 mm is recommended for the sludge discharge piping. Figure 5.22 illustrates a sludge sampling and withdrawal system in UASB reactors.

5.2.5 Sludge production and treatment

The solids accumulation rate depends essentially on the type of effluent being treated and is greater when the wastewater has a higher concentration of suspended solids, especially non-biodegradable solids.

In the case of treating soluble effluents, the production of excess sludge is very low and generally few problems are found in the handling, storage and disposal of the sludge. As a result of the low production and the high concentrations of

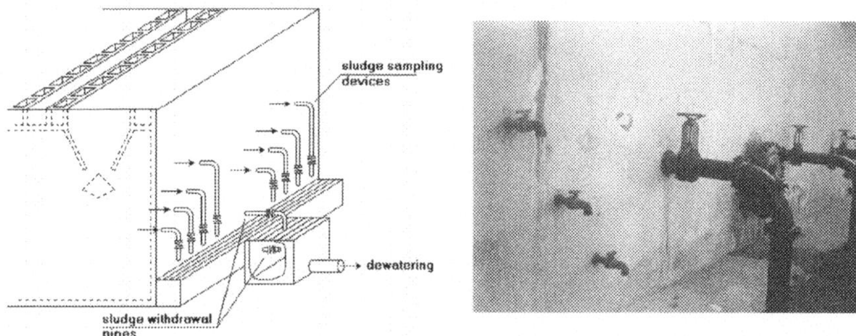

Figure 5.22.   Example of sludge sampling and discharge devices in UASB reactors

sludge in the reactor, the discharged volumes are also very small in comparison with aerobic processes.

Some important characteristics of anaerobic excess sludges from UASB reactors are as follows:

- high level of stability due to the high solids retention time in the treatment system, which allows the sludge to be directed to dewatering units without any prior treatment stage
- high concentration, usually in the order of 3 to 5%, allowing the discharge of smaller volumes of sludge
- good dewaterability
- possibility of the use of the dewatered sludge as a soil conditioner in agriculture, as long as care is taken regarding the presence of pathogens

(a)   Sludge production

The estimation of the mass production of sludge in UASB reactors can be done through the following equation:

$$P_s = Y \times COD_{app}$$

(5.26)

where:

P_s = production of solids in the system (kgTSS/d)
Y = yield or solids production coefficient (kgTSS/kgCOD$_{app}$)
COD_{app} = COD load applied to the system (kgCOD/d)

The values of Y reported for the anaerobic treatment of domestic sewage are in the order of **0.10 to 0.20 kgTSS/kgCOD$_{app}$**.

The estimation of the volumetric sludge production can be done by the following equation:

$$V_s = \frac{P_s}{\gamma \times (C_s/100)} \qquad (5.27)$$

where:

V_s = volumetric sludge production (m^3/d)
γ = sludge density (usually in the order of 1,020 to 1,040 kg/m^3)
C_s = solids concentration in the sludge (%)

(b)   Sludge dewatering

Sludge drying beds have been the alternative most commonly used for the dewatering of sludges from UASB reactors treating domestic sewage. This is due to the small volumes of sludge that are discharged from the system, as a result of the low yield and high concentration of the sludge in the reactors.

According to van Haandel and Lettinga (1984), the following procedures are necessary in the design of drying beds:

- estimate the daily mass of sludge that should be discharged from the reactor (Equation 5.26)
- define the productivity of the drying bed, to be expressed in terms of mass of solids that can be applied daily per unit area of the bed (kgTSS/m^2·d)
- adopt a maximum value of the fraction of the mass of sludge that can be discharged in one batch. Usually this fraction is in the order of 20 to 25% of the mass of sludge present in the reactor
- determine the sludge discharge frequency
- determine the number of beds

(c)   Wastewater pre-treatment

According to what was presented in Chapter 4, high-rate anaerobic reactors are designed with much smaller volumes in comparison with those of conventional anaerobic systems. For this reason, the entrance of non-biodegradable solids in the system is highly detrimental to the treatment process. The accumulation of this material in the reactor leads to the formation of dead zones and short circuits, significantly reducing the volume of biomass in the system and the efficiency of the treatment process.

Hence, the treatment of domestic sewage through high-rate anaerobic reactors is only possible if the flowsheet of the treatment plant incorporates preliminary treatment units (screens and grit chambers) aiming at the removal of coarse solids and inorganic settleable solids present in the sewage. In more recent designs a concern with the incorporation of devices that guarantee a greater efficiency in the removal of fine solids (that pass through conventional screens) and fats has been observed, aiming at guaranteeing better operational conditions in the reactor.

For example, the provision of sieves (static or mechanised) with openings in the order of 2 to 6 mm minimise the entrance of solids into the reactor, improving the functioning of the influent distribution device, due to the reduction/elimination of obstructions in the feeding tubes.

Regarding the provision of devices for the removal of fats, this is meant to reduce the scum formation problems in the reactor (as much in the interior of the gas collector as in the settler compartment). Scum, in fact, has frequently led to many operational problems due to the inherent difficulties in its removal.

5.2.6 Summary of the design criteria and parameters

A summary of the main criteria and parameters that orientate the design of UASB reactors for the treatment of domestic sewage, according to the previous items, is presented in Tables 5.9 and 5.10.

5.2.7 Construction aspects

(a)   Reactor height

The height to be adopted for the UASB reactors is dependent on the following main factors: (i) type of sludge present in the reactor; (ii) organic loads applied; and/or (iii) volumetric hydraulic loads, that define the upflow velocities imposed to the system. In the case of domestic sewage treatment in reactors that predominantly develop flocculent-type sludge, the upflow velocities imposed to the system lead to reactors with useful heights between 4.0 and 5.0 m, distributed in following way:

- height of settler compartment: 1.5 to 2.0 m
- height of digestion compartment: 2.5 to 3.5 m

Table 5.9. Summary of the main hydraulic criteria for the design of UASB reactors treating domestic sewage

Criterion/parameter	Range of values, as a function of flow		
	for Q_{ave}	for Q_{max}	for Q_{peak}[*]
Hydraulic volumetric load (m^3/m^3·d)	<4.0	<6.0	<7.0
Hydraulic detention time (hour)**	6 to 9	4 to 6	>3.5 to 4
Upflow velocity (m/hour)	0.5 to 0.7	<0.9 to 1.1	<1.5
Velocity in the apertures to the settler (m/hour)	<2.0 to 2.3	<4.0 to 4.2	<5.5 to 6.0
Surface loading rate in the settler (m/hour)	0.6 to 0.8	<1.2	<1.6
Hydraulic detention time in the settler (hour)	1.5 to 2.0	>1.0	>0.6

(*) Flow peaks with duration between 2 and 4 hours
(**) Sewage temperature in the range of 20 to 26 °C

Table 5.10. Other design criteria for UASB reactors treating domestic sewage

Criterion/parameter	Range of values
Influent distribution	–
Diameter of the influent distribution tube (mm)	75 to 100
Diameter of the distribution tube exit mouth (mm)	40 to 50
Distance between the top of the distribution tube and the water level in the settler (m)	0.20 to 0.30
Distance between the exit mouth and the bottom of the reactor (m)	0.10 to 0.15
Influence area of each distribution tube (m^2)	2.0 to 3.0
Biogas collector	–
Minimum biogas release rate (m^3/m^2·hour)	1.0
Maximum biogas release rate (m^3/m^2·hour)	3.0 to 5.0
Methane concentration in the biogas (%)	70 to 80
Settler compartment	–
Overlap of the gas deflectors in relation to the opening for the settler compartment (m)	0.10 to 0.15
Minimum slope of the settler walls (°)	45
Optimum slope of the settler walls (°)	50 to 60
Depth of the settler compartment (m)	1.5 to 2.0
Effluent collector	–
Submergence of the scum baffle or the perforated collection tube (m)	0.20 to 0.30
Number of triangular weirs (units/m^2 of the reactor)	1 to 2
Production and sampling of the sludge	–
Solids production yield ($kgTSS/kgCOD_{applied}$)	0.10 to 0.20
Solids production yield, in terms of COD ($kgCOD_{sludge}/kgCOD_{applied}$)	0.11 to 0.23
Expected solids concentration in the excess sludge (%)	2 to 5
Sludge density (kg/m^3)	1020 to 1040
Diameter of the sludge discharge pipes (mm)	100 to 150
Diameter of the sludge sampling pipes (mm)	25 to 50

(b)   Construction materials

Considering that the anaerobic degradation of certain compounds can lead to the formation of highly aggressive by-products, the materials used in the construction of anaerobic reactors should be resistant to corrosion.

For construction and cost reasons, concrete and steel have been the materials most commonly used in UASB reactors usually with an internal coating protection in an epoxy base. However, the solids and gas separator located in the upper part of the reactor that is more exposed to corrosion should be fabricated of a more resistant material or more heavily coated. Concrete is the material most frequently used, but experiences have not always been satisfactory due to problems of leaking gases, corrosion and that of constructing a bulky and heavy structure. Non-corrosive and less bulky materials such as PVC, fibreglass and stainless steel are more attractive options.

(c)   Corrosion protection

Resistance to corrosion can be intrinsic to the material (e.g. PVC, fibreglass, stainless steel) or can be part of it through special additives or coating/linings (e.g.

Table 5.11. Concrete coatings (comparative characteristics)

Coating	Advantage	Disadvantage
Chlorinated rubber	• Lower cost	• Lower resistance to volatile fatty acids
Bituminous epoxy	• Good resistance to volatile fatty acids • Can be applied with a greater thickness and a lower number of layers • Presents lower permeability	• Much higher cost

Source: Chernicharo *et al.* (1999)

concrete, steel). In the case of steel reactors, the care needs to be greater to avoid corrosion, including the use of special steels and the rigorous control of the coatings employed.

In the case of reinforced concrete reactors, the concern with the protection of the structure should be prior to the construction of the unit, such as in the provision of a concrete with sufficient chemical resistance. In this sense, some factors should be considered with the aim of obtaining lower rates of absorption and permeability:

- use of a concrete with a low water–cement ratio
- rigorous vibration of the concrete
- adequate curing process
- selection of an appropriate cement (Portland Pozzolanic)

In addition, the corrosion effects can be improved or inhibited through the application of acid resistant coatings. A thorough study in relation to the advantages of the different types of coatings was developed by Fortunato *et al.* (1998), that recommended possible coating solutions such as the painting of the reactor with chlorinated rubber or bituminous epoxy. These materials function as chemical barriers for the concrete surfaces exposed to highly aggressive environments. Some comparative characteristics of these types of coatings are presented in Table 5.11.

Example 5.2

Design a UASB reactor, based on the following design elements:

Data:

- Population: $P = 20,000$ inhabitants
- Average influent flow: $Q_{av} = 3,000$ m^3/d (125 m^3/hour)
- Maximum hourly influent flow: $Q_{max-h} = 5,400$ m^3/d (225 m^3/hour)
- Average influent COD (S_o) = 600 mg/L
- Average influent BOD (S_o) = 350 mg/L
- Sewage temperature: $T = 23$ °C (average of the coldest month)
- Solids yield coefficient: $Y = 0.18$ kgTSS/kgCOD$_{app}$
- Solids yield coefficient, in terms of COD: $Y_{obs} = 0.21$ kgCOD$_{sludge}$/kgCOD$_{app}$

Example 5.2 (Continued)

- Expected concentration of the discharge sludge: $C = 4\%$
- Sludge density: $\gamma = 1{,}020$ kg/m^3

Solution:

(a)   Calculation of the average influent COD load (L_o)

$$L_o = S_o \times Q_{av} = 0.600 \text{ kg/m}^3 \times 3{,}000 \text{ m}^3/\text{d} = 1{,}800 \text{ kgCOD/d}$$

(b)   Adopt a value for the hydraulic detention time (t)

$$t = 8.0 \text{ hours (according to Table 5.9)}$$

(c)   Determine the total volume of the reactor (V)

$$V = Q_{av} \times t = 125 \text{ m}^3/\text{hour} \times 8 \text{ hours} = 1{,}000 \text{ m}^3$$

(d)   Adopt the number of reactor modules (N)

$$N = 2$$

Although there is no limitation to the volume of the reactor, it is recommended that the reactor volume does not exceed 1,500 m^3, due to constructive and operational limitations. In the case of small systems for the treatment of domestic sewage, the adoption of modular reactors presents numerous advantages. In these cases, it has been usual to use modules with volumes no greater than 400 to 500 m^3.

(e)   Volume of each module (V_u)

$$V_u = V/N = 1{,}000 \text{ m}^3/2 = 500 \text{ m}^3$$

(f)   Adopt a value for the height of the reactor (H)

$$H = 4.5 \text{ m}$$

(g)   Determine the area of each module (A)

$$A = V_u/H = 500 \text{ m}^3/4.5 \text{ m} = 111.1 \text{ m}^2$$

Adopt rectangular reactors of 7.45 m $\times$ 15.00 m ($A = 111.8$ m^2)

(h)   Verification of the corrected area, volume and detention time

Corrected total area: $A_t = N \times A = 2 \times 111.8 \text{ m}^2 = 223.6 \text{ m}^2$
Corrected total volume: $V_t = A_t \times H = 223.6 \text{ m}^2 \times 4.5 \text{ m} = 1{,}006 \text{ m}^3$
Corrected hydraulic detention time: $t = V_t/Q_{av} = 1{,}006 \text{ m}^3/(125 \text{ m}^3/\text{hour}) = 8.0$ hours

(i)   Verification of the loads applied

Volumetric hydraulic load (Equation 5.8): $VHL = Q/V = (3{,}000 \text{ m}^3/\text{d})/1{,}006 \text{ m}^3 = 2.98 \text{ m}^3/\text{m}^3 \cdot \text{d}$

Example 5.2 (Continued)

Volumetric organic load (Equation 5.12): $L_v = Q_{av} \times S_o/V = (3{,}000 \text{ m}^3/\text{d} \times 0.600 \text{ kgCOD/m}^3)/1006 \text{ m}^3 = 1.79 \text{ kgCOD/m}^3 \cdot \text{d}$

(j)   Verification of the upflow velocities (according to Equation 5.15)

- for Q_{av}: $v = Q_{av}/A = (125 \text{ m}^3/\text{hour})/223.6 \text{ m}^2 = 0.56 \text{ m/hour}$
- for $Q_{max\text{-}h}$: $v = (225 \text{ m}^3/\text{hour})/223.6 \text{ m}^2 = 1.01 \text{ m/hour}$

It can be seen that the upflow velocities found are in agreement with the values shown in Table 5.9.

(k)   Influent wastewater distribution system

- Number of distribution tubes
  Adopting an influence area of 2.25 m² per distribution tube (according to Table 5.5), then the number of tubes can be calculated in accordance with Equation 5.21:
  $N_d = A/A_d = 223 \text{ m}^2/2.25 \text{ m}^2 = 99$ distributors. Due to the necessary symmetry of the reactor, adopt 100 distributors, as follows:
- along the length of each module (15.00 m): 10 tubes
- along the width of each module (7.45 m): 5 tubes

Thus, each module will have 50 (10 × 5) distribution tubes, each with an influence area equivalent to: $A_d = 223.6 \text{ m}^2/100 = 2.24 \text{ m}^2$.

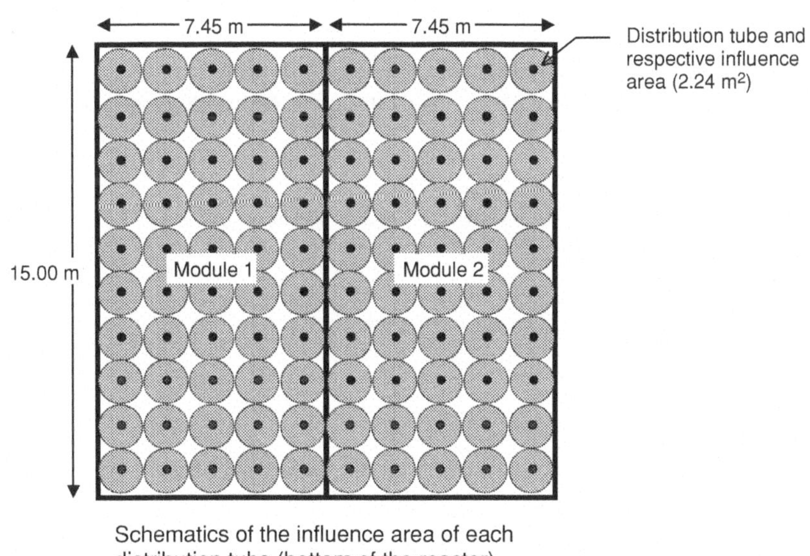

Schematics of the influence area of each distribution tube (bottom of the reactor)

Example 5.2 (Continued)

(l)   Estimation of the COD removal efficiency of the system

According to Equation 5.17:

$$E_{COD} = 100 \times (1 - 0.68 \times t^{-0.35}) = 100 \times (1 - 0.68 \times 8.0^{-0.35})$$
$$E_{COD} = 67\%$$

(m)   Estimation of the BOD removal efficiency of the system

According to Equation 5.18:

$$E_{BOD} = 100 \times (1 - 0.70 \times t^{-0.50}) = 100 \times (1 - 0.70 \times 8.0^{-0.50})$$
$$E_{BOD} = 75\%$$

(n)   Estimation of the COD and BOD concentrations in the final effluent

According to Equation 5.19:

$$C_{effl} = S_o - (E \times S_o)/100$$
$$C\ effl_{COD} = 600 - (67 \times 600)/100 = 198 \text{ mgCOD/L}$$
$$C\ effl_{BOD} = 350 - (75 \times 350)/100 = 88 \text{ mgBOD/L}$$

(o)   Evaluation of the methane production

The theoretical production of methane can be estimated from Equations 5.23, 5.24 and 5.25:

$$COD_{CH_4} = Q_{av} \times [(S_0 - C_{effl}) - Y_{obs} \times S_o)]$$
$$COD_{CH_4} = 3{,}000 \text{ m}^3/\text{d} \times [(0.600 - 0.198 \text{ kgCOD/m}^3)$$
$$- (0.21 \text{ kgCOD}_{sludge}/\text{kgCOD}_{app} \times 0.600 \text{ kgCOD/m}^3)]$$
$$COD_{CH_4} = 828 \text{ kgCOD/d}$$

$$K(t) = (P \times K_{COD})/[R \times (273 + T)]$$
$$K(t) = (1 \text{ atm} \times 64 \text{ gCOD/moL})/[(0.08206 \text{ atm·l/mol·K} \times (273 + 23 \text{ °C})]$$
$$K(t) = 2.63 \text{ kgCOD/m}^3$$

$$Q_{CH_4} = COD_{CH_4}/K(t)$$
$$Q_{CH_4} = (828 \text{ kgCOD/d})/(2.63 \text{ kgCOD/m}^3)$$
$$Q_{CH_4} = 314 \text{ m}^3/\text{d}$$

(p)   Evaluation of the biogas production

The evaluation of the biogas is done from the estimation of the percentage of methane in the biogas. Adopting a methane content of 75%:

$$Q_g = Q_{CH_4}/0.75 = (314 \text{ m}^3/\text{d})/0.75 = 419 \text{ m}^3/\text{d}$$

** *Example 5.2 (Continued)* **

(q) Sizing of the gas collectors

Number of gas collectors: 10 (5 in each module)
Length of each collector: $L_g = 7.45$ m (along the width of the reactor)
Total length of the gas collector: $L_t = 10 \times 7.45$ m $= 74.5$ m
Width of the upper part of the gas collector: $W_g = 0.25$ m (adopted)
Total area of the gas collectors (in its upper part): $A_g = L_t \times W_g = 74.5$ m $\times$ 0.25 m $= 18.6$ m^2
Verification of the biogas release rate in the gas collectors (K_g), according to Equation 5.22:

$$K_g = Q_g/A_i = (419/24 \text{ m}^3/\text{hour})/18.6 \text{ m}^2 = 0.94 \text{ m}^3/\text{m}^2 \cdot \text{hour}$$

Therefore, each biogas collector will have the following dimensions:

- Length: 7.45 m, Width: 0.25 m

(r) Sizing of the apertures to the settler compartment

Adopting 5 three-phase separators in each reactor, as illustrated below, then:
Number of simple apertures: 4 (2 in each module, alongside the walls)
Number of double apertures 8 (4 in each module, between the tri-phase separators)
Equivalent number of simple apertures: $4 + 8 \times 2 = 20$
Length of each aperture: $L_a = 7.45$ m (along the width of the reactor)
Equivalent length of simple openings: $L_t = 20 \times 7.45$ m $= 149.0$ m
Width of each aperture: $W_a = 0.40$ m (adopted)
Total area of the apertures: $A_t = L_t \times W_a = 149.0$ m $\times$ 0.40 m $= 59.6$ m^2
Verification of the velocities through the apertures (v_a):

- for Q_{av}: $v_a = Q_{av}/A_t = (125 \text{ m}^3/\text{hour})/59.6 \text{ m}^2 = 2.1$ m/hour
- for Q_{max-h}: $v_a = (225 \text{ m}^3/\text{hour})/59.6 \text{ m}^2 = 3.79$ m/hour

It can be seen that the velocities found are in agreement with the values in Table 5.9.
Therefore, each aperture to the settler compartment will have the following dimensions:

- Simple aperture: Length $= 7.45$ m, Width $= 0.40$ m
- Double aperture: Length $= 7.45$ m, Width $= 0.80$ m

(s) Sizing of the settler compartment

Number of settler compartments: 10 (5 in each reactor)
Length of each settler: $L_s = 7.45$ m (along the width of the reactor)

Example 5.2 (Continued)

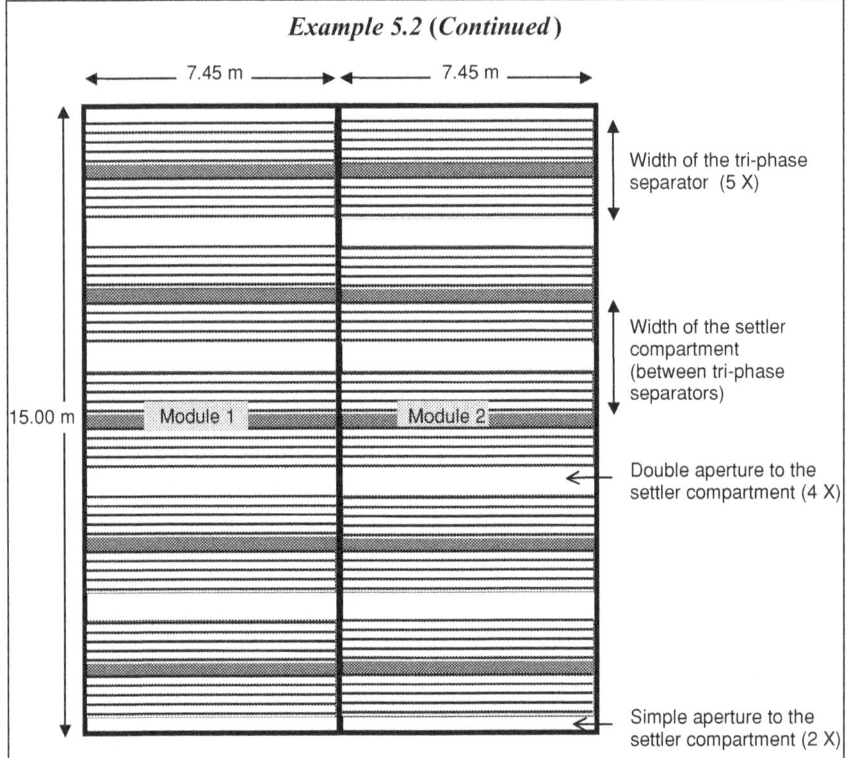

Schematics of the tri-phase separator distribution
(top of the reactor)

Total length of the settlers: $L_t = 10 \times 7.45\ m = 74.5\ m$
Width of each gas collector: $W_g = 0.30\ m$ (0.25 m + 0.05 wall thickness)
Width of each settler compartment: $W_s = 15.00\ m\ /\ 5 = 3.00\ m$
Effective width of each settler: $W_e = 3.00\ m - 0.30\ m = 2.70\ m$
Total area of the settlers: $A_s = L_t \times W_e = 74.5\ m \times 2.70\ m = 201.2\ m^2$
Verification of the surface loading rates of the settlers (v_s)

–   for Q_{av}: $v_s = Q_{av}/A_s = (125\ m^3/hour)/201.2\ m^2 = 0.62\ m/hour$
–   for $Q_{max\text{-}h}$: $v_s = (225\ m^3/hour)/201.2\ m^2 = 1.12\ m/hour$

It can be seen that the surface loading rates are in agreement with the values in Table 5.9. Therefore, each settler compartment will have the following dimensions in plan:

–   Length: 7.45 m, Width 2.70 m

Example 5.2 (Continued)

To determine the volume of the settler compartment, it is necessary to produce a general arrangement of the three-phase separator, taking into consideration the following aspects:

- height of the upper part of the settler compartment (vertical walls)
- height of the bottom part of the settler compartment (inclined walls)
- detention time for the settler compartment, in agreement with Table 5.9

(t)   Evaluation of the sludge production

The expected sludge production in the treatment system can be estimated from Equations 5.26 and 5.27

$$P_s = Y \times COD_{app} = 0.18 \; kgTSS/kgCOD_{app} \times 1{,}800 \; kgCOD/d$$
$$P_s = 324 \; kgTSS/d$$
$$V_s = P_s/(\gamma \times C_s) = (324 \; kgTSS/d)/(1{,}020 \; kg/m^3 \times 0.04)$$
$$V_s = 7.9 \; m^3/d$$

6

Operational control of anaerobic reactors

6.1 IMPORTANCE OF OPERATIONAL CONTROL

6.1.1 Preliminaries

The benefits of any wastewater treatment system, should it be either aerobic or anaerobic, will only be reached in an optimised manner if a logical sequence of actions is followed, as illustrated in Figure 6.1.

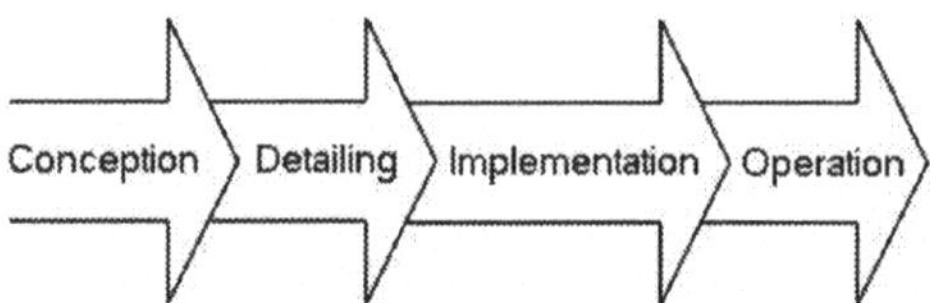

Figure 6.1.   Flowchart of actions for a wastewater treatment system

It is assumed from the above flowchart that the main objectives of any wastewater treatment system, that is, protection of the population's health and preservation of the environment, will only be achieved if the treatment plant is well conceived, well detailed, well implemented and also correctly operated. It is in relation to this last action that the operational control of the treatment plants becomes very important. Some aspects that demonstrate the relevance of the operational control are described in the following items, as highlighted in the original work developed by Chernicharo *et al.* (1999).

6.1.2 Verification of operational parameters

In countries with little tradition in wastewater treatment, the new treatment plants are usually designed based on parameters not always reliable and, many times, imported from foreign references. In general, these parameters can be verified during the operational phase of the system, taking into consideration the values originally assumed during the design phase. The various parameters of importance that should be verified during the operational phase of the system include:

- influent flowrates
- physical–chemical and microbiological characteristics of the influent wastewater
- efficiency and operational problems of the preliminary treatment units
- production and characteristics of the material retained in the screens and in the grit chamber
- efficiency and operational problems of the anaerobic reactor
- amount and characteristics of the biogas produced in the anaerobic reactor
- amount and characteristics of the sludge produced in the anaerobic reactor, etc.

These operational parameters, amongst others, can be properly evaluated based on the implementation of a monitoring programme, and later compared with the values originally assumed in the design, allowing:

- a revision or adaptation of the operational strategies initially planned for the treatment system
- a better based decision making, taking into consideration possible expansion plans for the system. Should the flows and organic loads be inferior to the design estimates, the project horizon can be increased and the investment with the expansion of the system in the subsequent years reduced. Conversely, flows and organic loads higher than those planned at the design stage will indicate that the project horizon should be reduced and that financial resources should be made available for the expansion of the system

6.1.3 Optimisation of the operational conditions

Another important aspect concerning the operational control of the treatment system is that it can lead to optimised operational conditions, aiming at reducing costs and meeting the discharge standards established by the environmental legislation. In this sense, some operational aspects should be emphasised:

- Determination of the best wastage and dewatering routine for the excess sludge. In the case of treatment plants that dewater the sludge on drying beds, wastage frequencies and solids loads to the beds leading to shorter drying cycles can be evaluated. Hence, an optimised sludge wastage and dewatering will directly imply a reduced volume of dry sludge to be transported to final disposal. An adequate wastage frequency will reflect directly on a smaller loss of solids in the final effluent, resulting in a better effluent quality in terms of suspended solids and particulate COD and BOD, with a direct impact on the compliance with the environmental legislation.

- Definition of the best practices and routines for operation and cleaning of the screening and grit removal units, aiming at optimising the efficiency of these preliminary treatment units. The removal of coarse materials and grit present in the influent wastewater can be maximised, preventing them from being introduced into the anaerobic reactor. These materials are highly harmful to the operation of the biological reactor, causing not only the obstruction of the sewage distribution piping, but also their accumulation inside the reactor, which causes the decrease of its useful volume and, consequently, a reduction in the efficiency of the system.

- Identification of bad odour points, aiming at providing a greater safety and environmental comfort to the operators and people who live near the treatment plant. In this sense, the effective follow-up of the units potentially subject to the release of foul gases (preliminary treatment, pumping station, anaerobic reactor and drying beds) will allow a greater knowledge of the problematic points, and facilitate the taking of measures and the implementation of adaptations to make odour control possible.

6.1.4 Workers' health and safety

In addition to the aspects previously mentioned, operational control is an important instrument for the identification of practices and routines that can promote the improvement of the workers' health and safety.

Health risks have always been a reason for concern in sewage treatment plants, since both *disability* and *occupational diseases* result in suffering and loss of human resources. Both cause a negative effect on the efficiency of the treatment system, on employees' morale, on public relationships and on costs (WEF, 1996). A good worker's health and safety programme should incorporate three main elements (USEPA, 1988; WEF, 1992):

- *Defined health and safety policy*: it comprises the principles of the whole health and safety programme, providing the workers with the key message of the programme, and making clear that it is supported by the upper management. The support should be visible, that is, the management should support the programme by means of actions and financial resources.

- *Work safety and occupational health committee*: it should be composed of management, supervisors and workers' representatives. Some specific tasks to be performed by the committee are: (i) conduct the health and safety programme; (ii) carry out systematic inspections; (iii) suggest and provide training; (iv) perform accident investigations; (v) maintain records on the occurrences; and (vi) prepare a health and safety manual.

- *Health and safety training*: the supervisors of the treatment plant should have, above all, their own attitudes and interests regarding health and safety, getting a total knowledge and understanding of the various forms of accident and occupational disease prevention. All new employees should undergo a health and safety programme, as well as all employees should be trained whenever a new equipment or process is added to the treatment plant.

Other details for the establishment of a health and safety programme for wastewater treatment plant operators can be found in WEF (1996).

6.2 OPERATIONAL CONTROL OF THE TREATMENT SYSTEM

6.2.1 Preliminaries

Although the operational simplicity of anaerobic treatment systems is one of its key points, the presence of operation and maintenance personnel is a necessary condition to assure appropriate performance. The three main treatment system control activities are:

- *operation*: refers to the daily or periodic activities necessary to assure a good and stable performance of the treatment system
- *maintenance*: refers to the activities to maintain the structures in the treatment plant in good conditions
- *information*: refers to the communication, preferably in writing, between the different people involved, creating, at the same time, a record of the operation and maintenance of the treatment system

6.2.2 Monitoring of the system

6.2.2.1 Need for system monitoring

The effective operational control of any wastewater treatment system will only be achieved by the implementation of an appropriate monitoring programme, to enable both the verification of the operational parameters and the optimisation of the operational routine.

The monitoring programme should be broad enough to include all the aspects relevant to the operation of the treatment system, without disregarding the local reality and the availability of human resources and material. Therefore, not only the development of physical–chemical and microbiological analyses becomes important, but also the gathering of a series of information on the operation of the system, as covered in the following items.

Usually, the anaerobic treatment systems can be divided into three parts, as presented in the schematic representation of Figure 6.2:

- pre-treatment
- biological treatment, or anaerobic digestion itself
- excess sludge dewatering

The operational activities of the anaerobic treatment systems are related to the different parts of the treatment system, and can be divided into four groups:

- Activities to ensure the appropriate operation of the pre-treatment units, usually consisting of: (i) screen (mechanised or not); (ii) grit chamber (mechanised or not); and (iii) flow measuring device, usually a Parshall flume coupled to the grit chamber
- Activities to evaluate the efficiency of digestion. Usually, anaerobic digestion is applied for the removal of suspended solids and organic matter, besides partially reducing the pathogenic organisms
- Activities to evaluate the operational stability of the digester, that is, to establish if there is any risk of the pH in the anaerobic reactor being reduced to a value lower than the minimum for the optimum methanogenesis ($pH_{min} = 6.5$)
- Activities to determine the amount and quality of the sludge in the reactor and in the excess sludge processing unit. The amount of sludge is important to establish the excess sludge wastage moment. The sludge quality is usually evaluated by specific methanogenic activity (SMA) and sedimentation tests. Regarding the quality of the excess sludge, the stability in which the sludge is wasted from the reactor and the solids fraction (or moisture fraction) in the dewatering unit (drying beds, centrifuges, filters or others) are important

In addition to these four specific groups, there may be others, depending on the intended use of the effluent. For instance, when the effluent is intended to be used (after a complementary treatment) in irrigation, it will be important to monitor the level of the nutrients N and P, although they do not play an important role in the treatment system and their removal is not the purpose of the anaerobic treatment.

In general, the tasks specified in the different groups will be carried out by different people. Thus, the works regarding the pre-treatment system require the frequent presence of personnel to verify whether there are blockages. Usually, the removal of coarse solids and sand collected in the pre-treatment units, as well as of dewatered sludge from the drying beds, will be manual, requiring unskilled labour. On the other hand, sampling of the biological treatment system and the undertaking of analyses to verify treatment efficiency, operational stability and the sludge mass in the reactor require more qualified personnel.

6.2.2.2 Monitoring programme

To facilitate the understanding of the units to be monitored in the system, Figure 6.2 presents a typical flowsheet of a sewage treatment plant consisting of the following units: preliminary treatment, anaerobic reactor and drying bed.

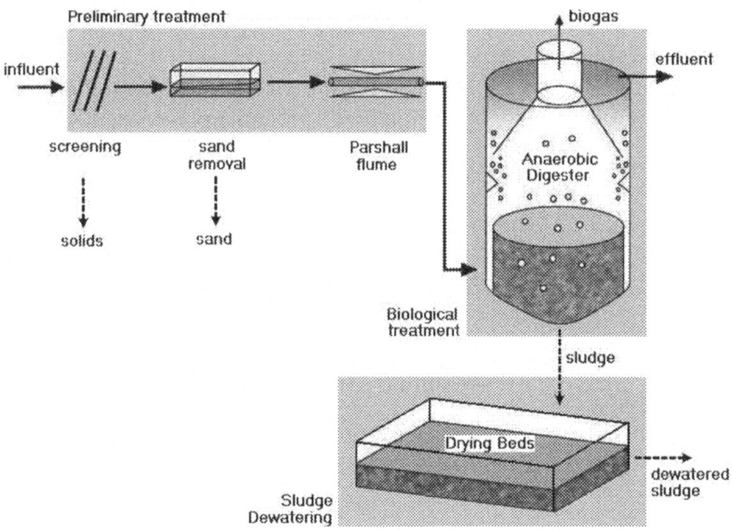

Figure 6.2.   Typical flowsheet of an anaerobic wastewater treatment plant with sludge dewatering and preliminary treatment units

This section focuses only on the operational control of reactors operating close to *steady-state conditions*, that is, the regime in which the system reaches more stable operational conditions, with no significant variations and instabilities over time. Recommendations on the operational control during the start-up period (*transient regime*) of the system are presented in Section 6.3.3.

(a)   Monitoring and operation of the preliminary treatment

Good operation of the anaerobic reactor depends fundamentally on the flow and characteristics of the wastewater to be treated and on the correct operation of the preliminary treatment units. An operational routine that allows the screens and grit chambers to be cleaned at a suitable frequency should be established to assure effective removal of the coarse solids and grit present in the wastewater. In the case of domestic sewage, the screen cleaning should be at least daily. Sand should be removed from the chambers once every 1 or 2 weeks, depending on the sand content in the influent wastewater (higher cleaning frequency for, say, 50 L of sand per 1,000 m^3 of influent sewage, and lower cleaning frequency for, say, 25 L of sand per 1,000 m^3 of influent sewage).

Regarding the most important characteristics that affect the anaerobic biodegradability (temperature and pH), these parameters can be easily measured in the influent. The preliminary treatment operation also includes the removal of blockages that may harm the uniform distribution of the influent in the treatment system. In this sense, the concentration of settleable solids is an important parameter.

The following figure and table (Preliminary treatment) identify the main points, parameters and frequency of monitoring at the preliminary treatment stage. The troubleshooting list presented at the end of this chapter identifies some problems that can be found in the daily operation of the preliminary treatment units.

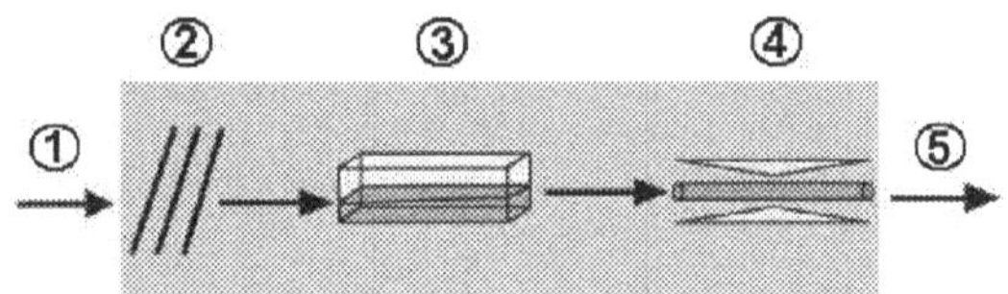

Preliminary treatment

		Monitoring points and frequency				
Parameter[1]	Unit	1	2	3	4	5
Flow	(L/s)	–	–	–	Daily	–
Screenings volume	(m³/d)	–	Daily	–	–	–
Sand volume	(m³/d)	–	–	Daily	–	–
Temperature	°C	–	–	–	–	Daily
pH	–	–	–	–	–	Daily
Settleable solids	(mL/L)	Daily	–	–	–	Daily

(b)   Monitoring of the anaerobic reactor

The successful operation of any anaerobic reactor depends on the systemisation and implementation of appropriate operational procedures during the start-up phase and over the operation on a steady-state basis. Three types of monitoring of the anaerobic reactor can be highlighted: (i) monitoring of the efficiency; (ii) monitoring of the stability; and (iii) monitoring of the amount and quality of the sludge.

Monitoring of the efficiency of the reactor

The historical behaviour of the unit and whether its performance is in accordance with the design specifications are established by monitoring the anaerobic reactor. Firstly, the course of the biological process itself is established, in terms of removal efficiency of undesirable constituents, by determining their concentrations in the influent and effluent of the biological reactor. The main constituents to be removed are:

- *suspended solids*: the concentration of suspended solids is determined by gravimetric tests on the total suspended solids (TSS) and on the volatile suspended solids (organic) (VSS). In addition, the traditional settleable solids test (determination of the volume of solids that settle in a 1-L cone during 1 hour) can be valuable if there is no precision scale available.
- *organic matter*: the organic matter removal efficiency is evaluated by the COD test and/or the BOD test. In addition, the biogas (or better, methane) production is a useful parameter in this respect.

- *pathogenic organisms*: regarding the hygienic quality, the establishment of the concentration of two types of organisms is recommended: (i) faecal coliforms (*E. coli*); and (ii) helminth eggs.

Monitoring of the stability of the digester

Monitoring of the operational stability of the treatment system aims at evaluating whether there are signs that the acid fermentation may prevail over the methanogenic fermentation, with the consequent acidification of the digester. In this sense, it is important to determine pH, alkalinity and concentration of volatile acids in the effluent, and compare these values with those in the influent. In addition, a sudden variation in the biogas composition and, especially, an increased percentage of carbon dioxide can be an indication of operational instability.

Monitoring of the sludge quantity and quality

Besides monitoring the efficiency and the stability of the reactor, tests should be performed to establish the quantitative and qualitative development of the sludge in the treatment system. The experimental determination of both presents problems. In systems with attached bacterial growth (immobilised biomass), such as fluidised bed reactors and anaerobic filters, the sludge is present in a form (biofilm) that makes its quantitative determination very difficult. In systems with dispersed bacterial growth, the concentration of sludge will not be uniform, and the determination of samples removed from several points is necessary. The concentration of both total solids (TS) and total volatile solids (TVS) should be determined.

The most important qualitative aspects of the sludge are:

- *Specific methanogenic activity*: reflects the capacity of the sludge to produce methane from an acetate substrate under optimised conditions. Although there are other processes developing in the anaerobic digester, the acetotrophic methanogenesis is the most important one because it is the limiting step in the conversion of the organic matter into methane. The test is performed in a laboratory according to the procedures described in Chapter 3. By knowing the SMA and the sludge mass in the biological reactor, it is possible to estimate the maximum organic load that can be digested in the reactor: this load is equal to the product of the SMA value and the sludge mass.
- *Stability*: aims at establishing which fraction of the sludge mass consists of still undigested biodegradable organic matter. A large fraction of biodegradable material in the sludge is not only an indication of an overloaded system, but it can also cause great problems to the solids–liquid separation of the excess sludge. Based on limited experience, van Haandel and Lettinga (1994) suggest that the fraction of biodegradable solids in the anaerobic sludge should be kept below 3%.
- *Settleability*: can be determined from a specific test described by Catunda and van Haandel (1989). This test is tedious and complicated, and the

application of a simpler, although less accurate method – the determination of the sludge volume index (SVI) or the diluted sludge volume index (DSVI) – is preferable in the operational routine.

The figure and table below (Anaerobic reactor) identify the main points, parameters and frequency of monitoring recommended for an anaerobic reactor. However, it should be highlighted that the monitoring parameters and frequency can be changed in view of local specificities and demands imposed by the environmental control agencies. A more intensive monitoring frequency may be necessary, particularly during the start-up of the system, as focused in the final items of this chapter.

(c)   Monitoring of the drying beds

As mentioned previously, optimised operational conditions of the sludge dewatering unit have direct implications on the reduction of the volume of dry sludge to be transported to the final disposal location and also on the quality of the effluent leaving the anaerobic reactor. Thus, to reduce the drying cycles of the excess sludge, a continuous monitoring of the solids should be performed inside the reactor (prior to wastage) and on the drying beds (after the wastage). This monitoring is essential to define the best sludge wastage and dewatering routine, to contribute to reduced drying cycles and to the attainment of a dry sludge with low water content.

The purpose of the sludge dewatering is to reduce the percentage of water in the sludge as much as possible and, at the same time, improve its hygienic quality, maintaining, as much as possible, the organic matter and the nutrients (nitrogen and phosphorus) in the most suitable form to turn the sludge into an organic fertiliser.

Regarding the operation of the drying beds, the most important parameters are:

- the load of solids applied to the bed
- the percolation time
- the composition and final quality of the dewatered sludge

Regarding the applied load, it is known that the sludge productivity (that is, the sludge mass that can be processed per unit area and per unit time to reach a certain desired final solids level) practically does not depend on the load applied, when it is within the range from 15 to 40 kg TS/m^2 (van Haandel and Lettinga, 1994).

Once the excess sludge is applied, the percolation and evaporation mechanisms start. As the fresh sludge flow is very small in relation to the sewage flow (approximately 0.1 to 0.2%), the composition of the percolated water (that returns to the treatment system) is not very important. The important parameters are the time necessary for percolation and the volume of percolated water. The water percolation is verified daily and, if applicable, the percolated volume is determined from the lowering of the sludge level on the bed (disregarding the water lost by evaporation).

Once percolation is finished, the composition in terms of total solids and the percentages of organic matter, nitrogen (organic and ammonia nitrogen) and phosphorus (total and orthophosphate) are determined at the end of the evaporation drying period.

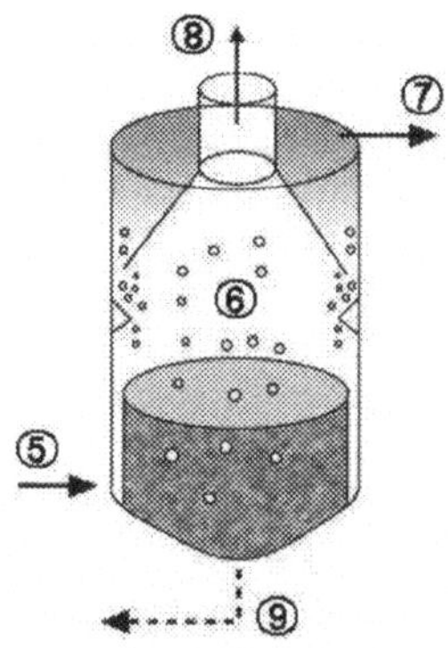

Anaerobic reactor

		Monitoring points and frequency				
Parameter	Unit	5	6	7	8	9
Treatment efficiency						
Settleable solids	mL/L	Daily	–	Daily	–	–
TSS	mg/L	Weekly	–	Weekly	–	–
COD	mg/L	Weekly	–	Weekly	–	–
BOD	mg/L	Monthly	–	Monthly	–	–
Biogas production	m^3/d	–	–	–	Daily	–
E. coli	MPN/100 mL	Weekly	–	Weekly	–	–
Helminth eggs[1]	N/L	Weekly	–	Weekly	–	–
Operational stability						
Temperature	°C	Daily	Daily	–	–	–
pH	–	Daily	Daily	–	–	–
Bicarbonate alkalinity	mg/L	Weekly	–	Weekly	–	–
Volatile fatty acids	mg/L	Weekly	–	Weekly	–	–
Biogas composition	$\%CO_2$	–	–	–	Monthly	–
Sludge quantity and quality						
Total solids[2]	mg/L	–	–	–	–	Weekly
Total volatile solids[2]	mg/L	–	–	–	–	Weekly
Specific methanogenic activity	gCOD/gVS·d	–	–	–	–	Monthly
Sludge stability	gCOD/gVS·d	–	–	–	–	Monthly
Sludge volume index (diluted)	mL/g	–	–	–	–	Monthly

Notes:

(1) The procedures for identification and enumeration of helminth eggs are described in the "Health guidelines for use of wastewater in agriculture and acquaculture". Technical Report Series (WHO, 1989) and in Zerbini and Chernicharo (2001).

(2) The analyses of total solids should be made at several points along the height of the bed and sludge blanket (3 to 6 points), to establish the profile and the mass of solids inside the reactor (see Chapter 3, Example 3.1)

Regarding the hygienic quality, it is convenient to determine the concentration of coliforms and viable helminth eggs. In practice, the end of the evaporation time will be usually established by the need to use the bed again to dry more sludge, or by the transport availability for the dry sludge.

The following figure and table (Sludge drying bed) identify the main points, parameters and frequency of monitoring recommended for drying beds.

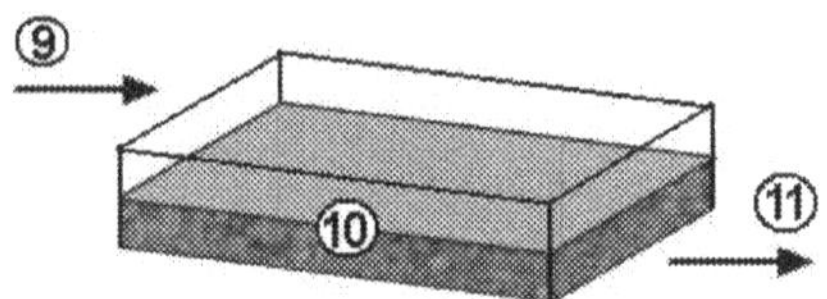

Sludge drying bed

Parameter[1]	Unit	Monitoring points and frequency		
		9	10	11
Excess sludge applied				
Volume of sludge wasted	m³	During wastage	–	–
Initial sludge concentration	gVS/L	During wastage	–	–
Initial sludge composition	%VS	During wastage	–	–
Applied load	kgTS/m²	–	During wastage	–
Faecal coliforms	MPN/gTS	During wastage	–	–
Helminth eggs	N/gTS	During wastage	–	–
Sludge on the bed and generation of percolate				
Height of the sludge	cm	–	During wastage	–
Percolation time	d	–	–	Daily[2]
Percolated volume	m³	–	–	Daily
Evaporation time	d	–	–	Daily
Solids percentage	%	–	2 × week	–
Sludge composition	%VS	–	2 × week	–
Total COD	mg/L	–	–	2 × week
Nitrogen (TKN)	% of the TS	–	[3]	–
Ammonia nitrogen	% of the TS	–	[3]	–
Total phosphorus (P_{tot})	% of the TS	–	[3]	–
Orthophosphate	$\%P_{tot}$	–	[3]	–
Faecal coliforms	MPN/gTS	–	[3]	–
Helminth eggs	N/gTS	–	[3]	–

Notes:
(1) The monitoring frequency refers just to the period between the sludge wastage from the reactor and the end of the drying period (removal of the dry sludge)
(2) Check daily whether there is water percolation
(3) The nitrogen, phosphorus and microbiological parameters should be analysed after the end of the drying period (in the dry sludge)

6.2.2.3 *Interpretation and recording of the operational data*

Some of the most important aspects of anaerobic systems are their simplicity and operational stability. In this sense, the operational database is more used for comparison between certain parameters and historical values. Corrective measures should be taken when a considerable, extended deterioration of the treatment performance is noticed.

The troubleshooting list at the end of this chapter identifies some problems and actions for their solution. These problems can only be identified when comparing the results of the different tests and analyses included in the monitoring programme with the values from previous periods. In many cases, the indication of an operational problem does not result from the absolute value of a parameter, but from its variation. Thus, the need to maintain frequent reports that characterise the performance and general situation of the treatment system is evident.

6.2.3 Sludge measurement and characterisation

(a)   Evaluation of the microbial mass

The determination of the biomass in anaerobic reactors presents two main difficulties:

- in some systems, the microorganisms are attached to small inert particles
- the biomass is usually present as a consortium of different morphological and physiological types

As presented in Chapter 3, the amount of biomass is usually evaluated by determining the solids profile, considering that the volatile solids are a measure of the biomass present in the reactors (mass of cellular material). The sludge samples are collected at different levels (heights) of the reactor, further gravimetrically analysed and the results are expressed in grams of volatile solids per litre (gVS/L). These concentration measures of volatile solids, multiplied by the volumes corresponding to each zone sampled, provide the masses of microorganisms over the profile of the reactor, as detailed in Example 3.1 (Chapter 3).

(b)   Evaluation of the microbial activity

The success of any anaerobic process, especially the high-rate ones, depends fundamentally on maintaining an adapted shock-resistant biomass with a high microbiological activity inside the reactor. So that this biomass is preserved and monitored, the development of techniques for the evaluation of the microbial activity of anaerobic reactors became imperative, especially of the methanogenic Archaea.

The SMA test can be used as a routine analysis to quantify the methanogenic activity of anaerobic sludge, or also in a series of other applications, as pointed out in Chapter 3, where detailed information is presented on the procedure for this test.

6.2.4 Wastage of sludge from the system

The accumulation of biological solids occurs in anaerobic reactors after some months of continuous operation. The solids accumulation rate depends essentially on the type of effluent being treated, being higher when the influent wastewater presents a high concentration of suspended solids, especially non-biodegradable ones. The accumulation of solids is also due to the presence of calcium carbonate or other mineral precipitates, besides the biomass production itself. When the accumulation of solids other than for bacterial growth prevails, it can be reduced by a pre-treatment (coagulation, flocculation, sedimentation). The accumulation of biomass depends essentially on the chemical composition of the wastewater, being higher for those with high carbohydrate concentrations.

(a)　Production of excess sludge and choice of the wastage point

To evaluate the amount of excess sludge produced in UASB reactors treating domestic sewage, an yield coefficient has been usually adopted ranging from 0.10 to 0.20 kgTSS per kgCOD applied to the system (see Chapter 5). In the case of reactor start-ups without seed sludge, the wastage of excess sludge should not be necessary during the initial months of operation of the reactor.

When the wastage is necessary in the case of UASB reactors, it should be done preferentially in the upper part of the sludge bed (less dense, more flocculent sludge, usually with lower specific methanogenic activity). However, attention should be given to the fact that the wastage of this lower concentration sludge will demand the removal of a larger sludge volume, for a given mass to be wasted, directly implying a larger area for the drying beds or a larger dewatering equipment.

An interesting alternative in relation to UASB reactors is to waste the sludge from different heights of the reactor, such as from the bottom (sludge bed) and from half-height of the digestion compartment (sludge blanket). Greater benefits can then be achieved than from the wastage from just a single height:

- The wastage from half-height of the digestion compartment enables the removal of the more disperse excess sludge, usually of lower activity and with poorer settleability.
- To compensate for the larger wastage volumes of this less dense sludge, a smaller portion of the sludge can be wasted from the bottom of the reactor, as it is very concentrated.
- The possible disadvantage to waste part of the bottom sludge, which usually presents higher activity and better settleability, can be compensated by the lower wastage volumes required and a consequent economy in the dewatering devices. Additionally, and depending on the quality of the preliminary treatment that precedes the reactor, the bottom sludge can accumulate inert solids, such as sand, which should be periodically discarded from the reactor. Consequently, the wastage of bottom sludge from the reactor, in small amounts and in a well-managed form, can bring important benefits to the treatment system.

(b)   Wastage of excess sludge

An important operational aspect in systems with dispersed growth, such as the UASB reactor, is the wastage of excess sludge. In this case, it is necessary that the sludge mass is maintained between a minimum (dictated by the need to have a sufficient treatment capacity in the system to digest the influent organic load) and a maximum (dependent on the sludge retention capacity of the system) value. The wastage of sludge together with the effluent should be minimised, since this wastage increases the concentration of COD, BOD and suspended solids in the effluent.

On the other hand, the wastage frequency will be dictated by the nature of the dewatering process. In case of a mechanical process, such as a centrifuge, the tendency will be for a daily wastage while the operator is present in the plant. Should there be a drying bed, the tendency will be to apply a large wastage, decreasing the sludge mass in the system from a value close to that of the maximum mass to a value a little higher than that of the minimum mass. Thus, the sludge wastage frequency is reduced to a minimum (and so is the work related to this wastage), while a good performance and operational stability of the digester are ensured. The following routine can be followed to establish the wastage frequency and magnitude (Chernicharo *et al.*, 1999):

- by operating the reactor under normal flow and load conditions, without discharging the excess sludge, the sludge mass in the reactor and the daily sludge production are determined for a reactor "full" of sludge
- the SMA of the sludge is determined
- from the SMA value, the minimum sludge required to maintain a good reactor performance is determined
- the difference between the maximum sludge mass that can be kept in the system and the minimum sludge mass necessary for a good reactor performance is calculated
- after a wastage equal to or lower than the maximum wastage, the loss of sludge together with the effluent is determined again
- the wastage frequency can be determined as the ratio between the sludge mass to be wasted and the sludge accumulation rate in the system

---

Example 6.1

Aiming at minimising the level of suspended solids in the effluent from a UASB reactor, estimate the wastage frequency of the excess sludge, assuming wastage of 50% of the sludge mass.

Data:

- total reactor volume: $V = 1{,}003.5 \ m^3$
- volume of the digestion compartment: $V_{dc} = 750.0 \ m^3$

Example 6.1 (Continued)

- volume of the sedimentation compartment: $V_{sc} = 253.5$ m^3
- depth of the reactor: 4.5 m
- average influent flowrate: $Q_{av} = 3{,}000$ m^3/d
- average influent COD concentration: $S_0 = 600$ mg/L
- average effluent COD concentration (in the absence of sludge wastage): $C_{eff} = 198$ mg/L
- average concentration of suspended solids in the effluent (in the absence of sludge wastage): 80 mg/L
- average effluent COD concentration (after sedimentation): 130 mg/L
- specific methanogenic activity of the sludge (at 24 °C): 0.34 mgCOD-CH$_4$/mgVS·d
- average effluent COD concentration (after wastage of 50% of the sludge mass): 140 mg/L
- average concentration of suspended solids in the effluent (after wastage of 50% of the sludge mass): 20 mg/L

Solution:

(a) Estimate the sludge mass when the reactor is full

Considering the data of Example 3.1, an estimate of 36,950 kgTS and 22,170 kgVS has been obtained (assuming an average fraction of volatile solids in the sludge equal to 60%).

(b) Estimate the sludge production in the system

The concentration of solids (that are considered sludge particles) in the effluent is equal to 80 mgTSS/L. Therefore, the daily sludge production is: 3,000 m^3/d × 0.080 kgTSS/m^3 = 240 kgTSS/d.

The volatile sludge concentration is estimated from the difference between the effluent (without wastage) and the settled effluent: $198 - 130 = 68$ mgCOD/L.

Knowing that 1 mgVS/L has a COD of 1.5 mgCOD/L, the volatile sludge concentration in the effluent is calculated as: (68 mgCOD/L)/(1.5 mg-COD/mgVS) = 45 mgVS/L.

Note that the specific sludge production, that is, the ratio between the daily sludge production (240 kgTSS/d) and the daily organic load applied $(3{,}000 × 0{,}600 = 1{,}800$ kgCOD/d) is equal to 0.13 kgTSS/kgCOD$_{applied}$, a value considered normal for anaerobic treatment.

(c) Estimate the sludge digestion capacity

From the specific methanogenic activity value and the volatile sludge mass, it is calculated that the sludge digestion capacity is: (0.34 kgCOD-CH$_4$/kgVS·d) × (22,170 kgVS) = 7,538 kgCOD/d.

Example 6.1 (Continued)

Note that the sludge digestion capacity is much higher than the influent load: $(3{,}000 \text{ m}^3/\text{d}) \times (0.600 \text{ kgCOD/m}^3) = 1{,}800 \text{ kgCOD/d}$.

(d)   Estimate the sludge accumulation in the reactor, after wastage

After wastage of 50% of the sludge, the loss of solids together with the effluent decreases to 20 mg/L, and the daily sludge production is reduced to: $3{,}000 \text{ m}^3/\text{d} \times 0{,}020 \text{ kgTSS/m}^3 = 60 \text{ kgTSS/d}$.

Therefore, the solids accumulation in the reactor can be estimated taking into account the sludge production before and after wastage: 240 kgTSS/d − 60 kgTSS/d = 180 kgTSS/d.

(e)   Estimate the wastage frequency of excess sludge

As the wastage of 50% of the maximum mass represents an amount of: 36,950 kgTSS $\times$ 0.50 = 18,475 kgTSS, it is estimated that a period of (18,475 kgTSS)/(180 kgTSS/d) = 102 days will be necessary to fill the reactor with sludge again.

Another approach is to say that the accumulation of 180 kgTSS/d represents an addition of (180 kgTSS/d)/(83.7 kgTSS/m^3) = 2.15 m^3/d in the lower part of the reactor (where the concentration is 50.2 gVS/L or 83.7 gTS/L, according to Example 3.1). Therefore, it can be considered that the monthly wastage rate would be 2.15 m^3/d $\times$ 30 d = 64.5 m^3 of the sludge from the bottom of the reactor.

Hence, wastage strategies of either 50% of the sludge every 102 days (which represents a volume of approximately 220 m^3 of the bottom of the reactor) or monthly 64.5 m^3 wastages, also from the bottom of the reactor, can be adopted. Alternatively, a more diluted sludge could be wasted in the upper areas, but then the wastage volume would be increased accordingly.

6.2.5 Prevention against the release of foul odours

Until recently, anaerobic processes were associated with foul odours, and this became the main barrier for their larger use in the treatment of liquid effluents. The large number of studies and researches being carried out in the area, notably from the 1970s, resulted in greater knowledge of the microbiology and biochemistry of the anaerobic process and, consequently, of the measures to be adopted for the control of these gases.

The formation of bad smelling gases is usually associated with the reduction of sulfur compounds to hydrogen sulfide (H_2S). Measures should be taken to prevent these gases from escaping to the atmosphere, especially when there are houses close to the treatment area. As the hydrogen sulfide can escape from the reactor both in the liquid (dissolved in the effluent) and in the gas (gas collector), different measures should be taken.

It is necessary to cover the reactor to prevent the H_2S dissolved in the effluent from escaping to the atmosphere. In this case, covering the reactor will also enable a reduced occurrence of corrosion, since the entrance of oxygen will be significantly reduced. The hydrogen sulfide that escapes from the reactor together with the effluent can be removed by some post-treatment method, such as chemical precipitation or chemical or biochemical oxidation. An important aspect to prevent the release of gases dissolved in the effluent relates to the design of the submerged collection system, to avoid turbulence (see Chapter 5).

In relation to the H_2S extracted by the gas collector, together with methane and carbon dioxide, there are some treatment alternatives that can be applied (Belli Filho *et al.*, 2001):

- adsorption, by the passage of the gas through a porous material, such as activated carbon
- absorption, by the contact between the gas and a slightly volatile liquid (solvent), for example in scrubbing towers. In these towers, the gas is applied against the current with the solvent, favouring the maximum contact between gas and liquid
- biological treatment, for example with biological filters and biofilters (for gases). In biological filters, the biogas flow passes through a scrubbing tower containing a high amount of biomass attached to a packing medium. Regarding the biofilters, the biogas is introduced into a tank containing biologically active material (compost) and the microorganisms undertake the reactions, generating innocuous products such as carbon dioxide, water, mineral salts and microbial biomass
- chemical precipitation, by the passage through a hydraulic seal containing some precipitating element, leading, for instance, to chemical precipitation of the sulfide as FeS

6.2.6 Other operational precautions

Besides the precautions previously mentioned, the operational routine of waste-water treatment plants should include other equally important aspects:

- verification and continuous cleaning of the feeding devices of the anaerobic reactors. This measure is particularly important in UASB-type reactors, as the correct wastewater distribution from the upper part to the lower part of the reactors is essential for the appropriate operation of the treatment unit. It is recommended that the wastewater distribution tubes are verified (and, if necessary, unobstructed) daily
- verification of the occurrence of corrosion in the structure of the anaerobic reactor, particularly in steel parts such as gas collectors, guard rails, etc. In case of occurrence of corrosion, the affected structures should be repaired

quickly, aiming at both the integrity of the treatment unit and the safety of the system operators

- correct destination of all solid materials removed in the preliminary treatment (screens and grit chamber) and sludge wasted from the anaerobic reactor
- removal of the floating material layer (scum) that tends to accumulate on the free surface of the sedimentation compartment and inside the gas collector.

6.3 START-UP OF ANAEROBIC REACTORS

6.3.1 Preliminaries

The reduction of the period necessary for the start-up and improved operational control of the anaerobic processes are important factors to increase the efficiency and the competitiveness of the high-rate anaerobic systems. However, a more critical discussion on the similarities, differences and advantages of the different high-rate anaerobic systems regarding start-up, operation and monitoring is difficult, once the behaviour of the process depends fundamentally on the characteristics of the wastewater to be treated.

In general, high-rate anaerobic processes can be operated with organic loads much higher than those of the conventional anaerobic reactors, but frequently these highly efficient processes require longer start-up periods, better operational control and more qualified operators, so that the maximum performance of the system is reached, with minimal risks of process failure. From the practical point of view, it is more economical to operate the reactor under lower loads, thus decreasing the efforts for the control of the operation and the process.

The start-up of the anaerobic reactors and, in a smaller scale, their operation has been considered by technicians as a barrier, possibly due to bad experiences linked to the use of unsuitable operational strategies. Therefore, systematised operational procedures are very important, mainly during the start-up of high-rate systems, notably in the case of UASB reactors.

The start-up of anaerobic reactors is determined by the initial transient period, marked by operational instabilities. The start-up can be basically achieved in three different manners:

- *by using seed sludge adapted to the wastewater to be treated*: the start-up of the system occurs fast, in a satisfactory way, as there is no need for acclimatisation of the sludge
- *by using seed sludge not adapted to the wastewater to be treated*: in this case, the start-up of the system goes through an acclimatisation period, including a microbial selection phase
- *with no use of seed sludge*: this is considered the most unfavourable form to start up the system, once it will be necessary to inoculate the reactor with its own microorganisms contained in the influent wastewater. As the concentration of microorganisms in the wastewater is very small, the time

required for the retention and selection of a large microbial mass can be very long (4 to 6 months)

The start-up and operation of anaerobic filters and UASB reactors are covered in the following items, with special emphasis to the latter ones.

6.3.2 Start-up and operation of anaerobic filters

Usually, the start-up of anaerobic filters for the treatment of domestic sewage has not received much attention, possibly due to the following main aspects:

- anaerobic filters have been primarily applied to the treatment of the sewage from small populations (frequently below 500 inhabitants), and they are not the object of larger operational care in view of the dimension of the systems
- these reactors are provided with a packing medium, ensuring a larger retention of solids and biomass in the system, favouring the start-up process.

However, the anaerobic filters can be started up similarly to the UASB reactors, that is: (i) without seed sludge; (ii) with seed sludge not adapted to the type of wastewater to be treated; and (iii) with seed sludge adapted to the type of wastewater. As such aspects are covered in more detail in the following section, where guidelines for the start-up of UASB reactors are presented, only the aspects inherent to the anaerobic filters are discussed here.

(a)   Grease removal

The problem of grease entering into a sewage treatment system results from the characteristics of this material, which tends to accumulate on the upper surface of the treatment units. As they are considered slow and hardly biodegradable materials, they form, together with other floating materials, a thick scum layer, which reduces the useful volume of the tank and tends to harm its operation.

The need for the implementation of grease removal units upstream the anaerobic filters depends intrinsically on the amount of oils and greases present in the wastewater. Although the implementation of these units is not a regular practice, the occurrence of operational problems due to the large presence of grease and the consequent scum formation in anaerobic reactors, particularly in the UASB reactors, has led several new designs of treatment plants to consider the implementation of a grease removal unit upstream the anaerobic reactors.

(b)   Coarse solids removal

Like any other sewage treatment system, it is essential that the anaerobic filter is preceded by a preliminary treatment unit intended for the removal of coarse solids. This unit may consist of a screen, or simply of a collecting basket, depending on the size of the system and on the amount of coarse material present in the sewage.

The non-incorporation of coarse solids removal units preceding anaerobic filters contributes negatively to the occurrence of operational problems in these units.

For example, when larger floating solids have access to an anaerobic filter, they can obstruct the holes of the upper slab of the bottom compartment of the filter, which is a problem difficult to correct. In certain situations, when plastic bags, condoms and other similar objects are retained in the bottom compartment, the correction of the problem may require the closure of the filter, the removal of the packing medium and the removal of the bottom slab, to withdraw the material that caused the obstruction. Thus, it is essential to install a screening unit or a collecting basket upstream the anaerobic filters.

Having in mind that the installation of a screening unit or collecting basket has a very low cost compared with the other units of the system, it is recommended that these units are always present in any sewage treatment system.

(c)   Wastage of sludge from the system

Young (1991) recommends that the solids should not be wasted from the reactor until the concentration in the sludge zone exceeds 5% (dry solids). Even in these conditions, wastage should only be performed if the sludge blanket penetrates the packing medium or if the concentration of solids in the effluent increases significantly. If the sludge blanket is not distinguished from the sludge bed (uniform distribution), solids should be wasted whenever the solids concentration is approximately 7%, in which case the flow of the solid mass will be hindered, which may favour the formation of preferential routes for the wastewater, besides hindering the removal of excess sludge.

6.3.3 Start-up and operation of UASB reactors

The successful application of the high-rate anaerobic processes is subject to the compliance with a series of requirements, which are mainly related to the concentration and activity of the present biomass, and also to the mixing and flow regime in the reactor, considering that all environmental factors (temperature, pH, alkalinity, etc.) are within the optimum range.

The most common objectives to be achieved in the operation of high-rate anaerobic processes are the control of the solids retention time (independently from the hydraulic detention time), the prevention against the accumulation of inert suspended solids in the reactor and the development of favourable conditions for mass transfer. These objectives are generally achieved when the reactors are well designed and constructed, and when appropriate procedures during the start-up and operation of the system are taken.

(a)   Grease removal

The same considerations made in the previous section for anaerobic filters, regarding the importance of the installation of grease removal units preceding anaerobic reactors, are valid for the UASB reactors. The operational problems resulting from the non-removal (or inadequate removal) of grease can be highly detrimental, as these materials may enable the excessive accumulation of scum inside the gas collectors, hindering the release of gases and demanding special devices for its periodical removal.

Although the installation of a grease removal unit upstream UASB reactors is not a regular practice yet, the operational problems that have occurred in units already installed have called the designers' attention towards the inclusion of this unit in the design of new treatment plants.

(b) Removal of coarse solids

As highlighted in the beginning of this chapter and in Chapter 5, the effective removal of coarse solids before the sewage is directed to the UASB reactors is essential. In the particular case of the UASB reactors, the operational problems resulting from the non-removal (or inadequate removal) of coarse solids may jeopardise the whole operation of the treatment system, once these materials can adversely affect the distribution of the influent wastewater at the bottom of the reactor, and generate and accumulate a sludge with poor characteristics, with low activity and difficult to remove.

The concern with the excessive entry of larger dimension solids in the UASB reactors is so great that many of the new designs have considered the installation of sieves, with openings from 1 to 5 mm, to reduce at the most the operational problems resulting from the entry of solids into the reactor.

(c) Considerations and criteria for the start-up of the system

Volume of seed sludge. The volume of seed sludge for the start-up of the system is usually established as a function of the initial biological loading rate applied to the treatment system. The biological loading rate (kgCOD/kgVS·d) is the parameter that characterises the organic load applied to the system in relation to the amount of biomass present in the reactor (see Chapter 5, Equation 5.14). The biological load values to be applied during the start-up depend essentially on the type of seed sludge employed and on its acclimatisation to the wastewater to be treated. It is recommended that whenever possible, the biological load for the start-up be established by means of specific methanogenic activity tests of the sludge (see Chapter 3). Should it be impossible to perform these tests, biological loads in the range from 0.10 to 0.50 kgCOD/kgVS·d, relating to specific methanogenic activities between 0.10 and 0.50 kgCOD-CH_4/kgVS·d, are used during the start-up of the process. These initial loads should be gradually increased according to the efficiency of the system and the improved activity of the biomass.

Volumetric hydraulic load. The volumetric hydraulic load is equal to the amount (volume) of sewage applied daily to the reactor per unit volume (see Chapter 5, Equation 5.8). The hydraulic load produces at least three different effects on the biomass of the reactor during the start-up of the system:

- the hydraulic load removes all the biomass with poor settling characteristics, thus creating space for the new biomass that is growing
- with the removal of part of the new biomass, which does not have good settleability, a selection of the active biomass is made
- the hydraulic load has a strong influence on the mixing characteristics of the reactor, mainly during the start-up of the system

In view of that, the dilution of very concentrated wastewater (COD > 5,000 mg/L) is essential, aiming at obtaining higher hydraulic loads during the initial transient period (Lettinga *et al.*, 1984).

Biogas production. Biogas production is very important in UASB reactors for good mixing of the sludge bed. However, very high gas production rates can adversely affect the start-up of the process because the sludge can expand excessively towards the upper part of the reactor, being lost together with the effluent.

Temperature. The ideal operation temperature for anaerobic reactors is in the range of 30 to 35 °C, when the growth of most of the anaerobic microorganisms is considered ideal. In the case of domestic sewage treatment, this range of temperature is hardly reached, once the average temperature of the influent sewage in warm-climate regions usually ranges from 20 to 28 °C. Under these sub-optimum temperature conditions, the anaerobic reactors are started up more easily with the inoculation of sufficient amounts of anaerobic sludge, preferably acclimatised to the type of sewage.

Environmental factors. For an optimum start-up of the system, it is desirable that the environmental factors are favourable, in accordance with the following main guidelines:

- whenever possible, the temperature inside the reactors should be close to the ideal growth range of anaerobic microorganisms (30 to 35 °C). In the case of domestic sewage treatment, these temperatures are not feasibly reached, which makes the start-up of the system under the ideal temperature conditions virtually impossible
- pH should be always maintained above 6.2 and, preferably, in the range from 6.8 to 7.2
- all the growth factors (N, P, S and micronutrients) should be present in sufficient amounts
- the toxic compounds should be absent in inhibiting concentrations. Otherwise, sufficient time should be provided for acclimatisation of the microorganisms

Acclimatisation and selection of biomass. The first start-up of an anaerobic reactor is a relatively delicate process. In the case of UASB reactors, sufficient, continuous removal of the lightest sludge fraction is essential, to allow the selection of the heaviest sludge for growth and aggregation. The main guidelines for acclimatisation and selection of biomass in UASB reactors are as follows (adapted from: Lettinga *et al.*, 1984):

- do not return to the reactor the dispersed sludge lost together with the effluent
- dilute the influent or recirculate the effluent, when the concentration of wastewater exceeds 5,000 mgCOD/L
- increase the organic load progressively, whenever the BOD or COD removal efficiency reaches at least 60%

- keep the acetic acid concentrations below 1,000 mg/L. In the case of domestic sewage treatment, the expected acetic acid concentrations in the reactor are much lower, and they should be maintained below 200 to 300 mg/L
- provide the necessary alkalinity to the system, to maintain the pH close to 7

(d)   Procedure preceding the start-up of a reactor

Characterisation of the seed sludge. Once the use of seed sludge is defined for the start-up of the reactor, analyses should be carried out for its qualitative and quantitative characterisation, including the following parameters: pH, bicarbonate alkalinity, volatile fatty acids, TS, VS, and SMA. Besides the parameters referred to above, a visual and olfactory characterisation of the sludge should be carried out.

Characterisation of the raw sewage. To establish the start-up routine of the anaerobic reactor, a qualitative and quantitative characterisation campaign of the influent raw sewage should be carried out.

Estimation of the seed sludge volume necessary for the start-up of the reactor. Based on the results of the characterisation analyses of the sludge and the influent sewage, the seed sludge volume necessary for the start-up of the reactor can be estimated, as shown in Example 6.2.

Example 6.2

Estimate the amount of sludge necessary for the inoculation of a UASB reactor, knowing the following elements:

Data:

- Influent flowrate: $Q_{av} = 3,000$ m^3/d (adopted as an average of the characterisation campaign)
- Sewage concentration: $S_o = 600$ mgCOD/L (adopted as an average of the characterisation campaign)
- Concentration of volatile solids in the seed sludge: $C = 30,000$ mgVS/L (3%) (adopted as an average of the samples analysed)
- Density of the seed sludge: $\gamma = 1,020$ kg/m^3
- Volume of the reactor: $V = 1,003.5$ m^3
- Biological loading rate adopted during the start-up of the reactor: $L_s = 0.3$ kgCOD/kgVS·d

Solution:

- Applied organic load (L_o):
$$L_o = Q_{av} \times S_o = 3,000 \text{ m}^3/\text{d} \times 0.600 \text{ kgCOD/m}^3$$
$$L_o = 1,800 \text{ kgCOD/d}$$

- Necessary seed sludge mass (M_s):
$$M_s = L_o/L_s = (1,800 \text{ kgCOD/d})/(0.3 \text{ kgCOD/kgVS·d})$$
$$M_s = 6,000 \text{ kgVS}$$

Example 6.2 (Continued)

- Resulting seed sludge volume (V_s):
  $V_s = P_s/(\gamma \times C_s)$ – see Chapter 5, Equation 5.27
  $V_s = (6{,}000 \text{ kgVS})/(1{,}020 \text{ kg/m}^3 \times 0.03)$
  $V_s = 196 \text{ m}^3$

As the necessary seed sludge volume is relatively high (196 m^3), equivalent to approximately 32 tank trucks, the possibility of not applying the total organic load can be evaluated, diverting (by-passing) part of the influent sewage to the overflow weir of the treatment plant during the first few days of the reactor start-up.

The following figure enables the visualisation of some alternatives for inoculation and start-up of the anaerobic reactor, taking into consideration the application of different influent flow percentages as a function of the volatile solids concentrations in the sludge.

In the figure, the percentage of applied flow refers to the average flow obtained in the characterisation campaign of the influent (e.g.: 50% refers to the application of an influent flowrate equal to 1,500 m^3/d). Possible alternatives for inoculation of the reactor can be evaluated by means of graphical aid, as exemplified below:

- for application of 100% of the influent flowrate, considering a sludge with a concentration of volatile solids equal to 3%, a seed sludge volume equal to approximately 200 m^3 is necessary
- for application of 50% of the influent flow, considering a sludge with a concentration of volatile solids equal to 5%, a seed sludge volume equal to approximately 60 m^3 is necessary

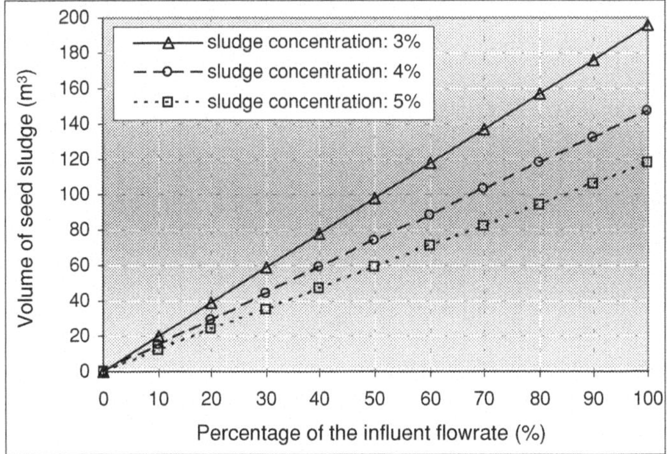

Graphical representation of seed sludge volumes necessary for the start-up of an UASB reactor, considering the conditions of Example 6.2 and different seed sludge concentrations

(e)   Procedure during the start-up of an anaerobic reactor

The procedure during the start-up of the reactor refers mainly to: (i) inoculation, (ii) feeding with wastewater and (iii) monitoring of the process.

Inoculation of the reactor

The inoculation can be done with the reactor either full or empty, although the inoculation is preferable with the reactor empty, to reduce sludge losses during the transfer process. For this second situation, the following procedures can be adopted:

- transfer the seed sludge to the reactor, ensuring that it is discharged into the bottom of the reactor. Avoid turbulence and excessive contact with air
- leave the sludge at rest for an approximate period of 12 to 24 hours, allowing its gradual adaptation to local temperature

Feeding of the reactor with sewage

- after the end of the rest period, begin the feeding of the reactor with wastewater, until it reaches approximately half of its useful volume
- leave the reactor unfed for a 24-hour period. At the end of this period, and prior to beginning the next feeding, collect supernatant samples from the reactor and analyse the following parameters: *temperature, pH, alkalinity, volatile acids and COD*. Should these parameters be within acceptable ranges, continue the feeding process. *Acceptable values: pH between 6.8 and 7.4 and volatile acids below 200 mg/L (as acetic acid)*
- continue the filling process of the reactor, until it reaches its total volume (level of the sedimentation tank weirs)
- leave the reactor unfed again for another 24-hour period. At the end of this period, collect new samples for analyses and proceed as previously stated
- if the parameters analysed are within the established ranges, feed the reactor continuously, in accordance with the amount of seed sludge used and the flow percentage to be applied (see above figure)
- implement and perform a routine monitoring of the treatment process
- increase the influent flow gradually, initially every 15 days, in accordance to the system response. This interval can be either increased or reduced, depending on the results obtained

Monitoring of the treatment process

For the monitoring of the treatment process, the sample collection routine and the physical–chemical parameters to be analysed should be defined during the start-up period. An example of a monitoring programme that has been adopted in the start-up of UASB reactors is presented in Table 6.1.

Table 6.1. Monitoring programme of a UASB reactor during the start-up period

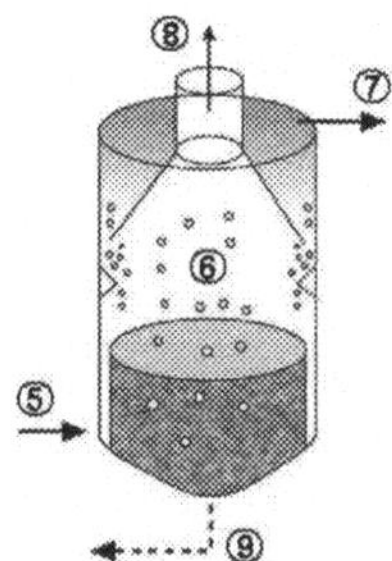

		Monitoring points and frequency[1]				
Parameter	Unit	5	6	7	8	9
Treatment efficiency						
Settleable solids	mL/L	Daily	–	Daily	–	–
TSS	mg/L	3 × week	–	3 × week	–	–
Total COD	mg/L	3 × week	–	3 × week	–	–
Total BOD	mg/L	Weekly	–	Weekly	–	–
Biogas production	m^3/d	–	–	–	Daily	–
Operational stability						
Temperature	°C	Daily	Daily	–	–	–
pH	–	Daily	Daily	–	–	–
Bicarbonate alkalinity	mg/L	3 × week	–	3 × week	–	–
Volatile fatty acids	mg/L	3 × week	–	3 × week	–	–
Biogas composition	$\%CO_2$	–	–	–	Weekly	–
Sludge quantity and quality						
Total solids[2]	mg/L	–	–	–	–	Monthly
Total volatile solids[2]	mg/L	–	–	–	–	Monthly
Specific methanogenic activity	gCOD/gVS·d	–	–	–	–	2 × month
Stability of the sludge	gCOD/gVS·d	–	–	–	–	Monthly
Sludge volume index (diluted)	mL/g	–	–	–	–	Monthly

Notes:

(1) The analysis frequency can be reduced over the start-up of the process, in accordance with the results achieved

(2) The total solids should be analysed at various points along the height of the bed and sludge blanket (3 to 6 points), to obtain the profile and the mass of solids inside the reactor (see Chapter 3, Example 3.1)

6.4 OPERATIONAL TROUBLESHOOTING

The following items present a set of information that can help detect and correct operational problems in anaerobic reactors, based on the work of Chernicharo *et al.* (1999).

Flow and characteristics of the influent

Observation	Probable cause	Verify	Solution
Flow always lower than the expected one	Population or *per capita* contribution lower than the design value	Flow measuring device	Increase served population
Flow suddenly lower than the expected one	Blockages in sewerage system	Overflow in the contribution area	Unblock sewers
Flow always higher than the expected one	Population or *per capita* contribution higher than the design value	Flow measuring device	Increase treatment capacity
Daily peaks higher than the expected ones	Equalisation lower than the expected one	Flow measuring device	Consider equalisation tank
Sudden irregular peaks	Combined system or cross-connection with stormwater sewers	Coincidence with rains	Disconnect illegal connections
Flow sometimes higher than the expected one	Large infiltration of groundwater	Coincidence with rains	Find the infiltration points
pH higher or lower than normal	Industrial wastewater	Existence of illegal sources	Find and act on the sources to correct the problem
Temperature higher or lower than the normal	Industrial waste	Existence of illegal sources	Find and act on the sources to correct the problem
Settleable solids larger than normal	Illegal dumping of domestic or industrial solid wastes in the sewerage system	Nature of the settleable solids	Find and act on the sources to correct the problem

Source: Chernicharo *et al.* (1999)

Preliminary treatment

Observation	Probable cause	Verify	Solution
Odour or insects at the screen	Long interval between cleanings	Cleaning interval	Increase the cleaning frequency
Sudden increase in the mass of coarse solids retained	Illegal dumping of solid wastes	Existence of illegal sources	Find and act on the sources to correct the problem
Sudden decrease in the mass of coarse solids retained	Retention failure at the screen	Condition of the screen	Repair the screen
Sudden increase in the grit mass retained	Discharge of stormwater into the sewerage system	Sewage flow	Disconnect illegal connection
Sudden decrease in the sand mass retained	Sand dragged from the grit chamber	Flow velocity (dye tracer)	Reduce velocity
Rotten egg odour in the grit chamber	Sedimentation of organic matter	Flow velocity (dye tracer)	Increase water velocity
Sand retained is grey, has a bad odour and contains grease	Sedimentation of organic matter	Flow velocity (dye tracer)	Increase water velocity
Metal and concrete corrosion in the preliminary treatment units	Insufficient ventilation	Ventilation	Improve ventilation

Source: Chernicharo *et al.* (1999)

Performance of the UASB reactor

Observation	Probable cause	Verify	Solution
Unequal influent distribution	Distribution structure unlevelled	Level of the distribution structure	Level the distribution structure
Distribution tube does not receive sewage	Blocking	Blocking	Unblock
Non-uniform effluent collection	Collection structure unlevelled	Level of the collection structure	Level the collection structure
	Surface layer obstructs collection points	Flow conditions	Remove obstruction
High level of settleable solids in the effluent	Excessive hydraulic load	Flow	Reduce flow
	Excessive solids in the reactor	Sludge mass	Waste the excess sludge
Gas production lower than normal	Biogas leakage	Gas collection	Eliminate leakage
	Defective gas meter	Gas meter	Either repair or replace
	Reduced flow	Influent flow	Unblock sewers
	Toxic material in the influent	SMA test	Identify and act on sources of toxic material
	Excessive organic load	SMA and stability test	Reduce organic load
Sludge production higher than normal	Overloaded sludge	Sludge stability	Reduce applied load
	Coarse and/or inorganic solids entering the reactor	Pre-treatment operation	Re-establish operation of the pre-treatment units
Sludge production lower than normal	Small flow	Influent flow	Unblock sewers
	Deficient sludge retention	Phase separator; settleable solids in the effluent	Repair separator
Sludge with high fraction of inorganic solids	Defective grit chamber	Velocity in the chamber	Decrease velocity in grit chamber
	Low upflow velocity in the reactor	Velocity	
Floating sludge grows quickly	Excessive hydraulic load	Organic and hydraulic loads	Reduce load
Reduced efficiency in the removal of organic matter	Excessive load	Load	Reduce load
	Deficient influent distribution	Influent distribution system (tracer studies)	Repair failure

Source: Chernicharo *et al.* (1999)

Characteristics of the sludge in the reactor

Observation	Probable cause	Verify	Solution
SMA lower than the expected one	Entry of inert solids	Settleable solids in influent	Reduce source or revise pre-treatment
	Overload	Sludge stability and removal efficiency of the organic matter	Reduce load
	Presence of toxic material	Test stored sludge	Identify and act on sources of toxic materials
Poor stability	Sludge overload	Specific organic load	Reduce specific load
High sludge volume index	Biodegradable organic matter	Stability	Reduce organic load
	Low hydraulic load	Upflow velocity	Increase dragging temporarily
Poor settleability	Dispersed flocs due to excessive organic load	Sludge stability	Reduce load
	Presence of toxic material	SMA of the sludge	Identify and act on sources of toxic materials
Increased specific sludge production	Flocculation without metabolism	Sludge stability	Reduce specific organic load
Increased inorganic fraction	Entrance of silt and sand	Velocity in the grit chamber	Reduce velocity in the grit chamber
	Low upflow velocity	Upward velocity in the reactor	Increase hydraulic load

Source: Chernicharo *et al.* (1999)

Sludge drying beds

Observation	Probable cause	Verify	Solution
Generation of bad odour when applying sludge to the bed	Sludge instability	Sludge stability (test)	Adjust organic load
Excess sludge wastage tubing blocked	Accumulation of solids and sand	Occurrence of blocked pipes	Clean tubing after use
Excessive percolation time	Excessive load applied	Applied load	Reduce load
	Inadequate bed cleaning	–	Improve maintenance
	"Blind" sand	Verify permeability	Replace sand
	High rainfall	–	Cover bed
	Drainage system blocked	–	Apply upflow washing
	Air trapped in the bed preventing passage of water	Upflow washing with water	Apply water in upward direction, saturating the bed before sludge wastage
Excessive evaporation time	Excessive load applied	Load applied	Reduce load
	High rainfall, low temperatures, high air humidity		Reduce load/cover bed
Very diluted excess sludge	Sludge removal from a very high level in the reactor	Solids concentration profile	Remove the sludge from a lower level (closer to the bottom of the reactor)
Mosquito reproduction on the beds	Semi-permanent water layer	Drainage system	Reduce load, improve permeability

Source: Chernicharo *et al.* (1999)

7

Post-treatment of effluents from anaerobic reactors

7.1 APPLICABILITY AND LIMITATIONS OF THE ANAEROBIC TECHNOLOGY

7.1.1 Applicability for the treatment of domestic sewage

A deep discussion on the evolution and applicability of the anaerobic technology for the treatment of domestic sewage was presented in Chapter 1, where several favourable characteristics of the anaerobic processes were highlighted, such as low cost, operational simplicity, no energy consumption and low production of solids. These advantages, associated with favourable environmental conditions in warm-climate regions where high temperatures prevail practically throughout the year, have contributed to establish the anaerobic systems, particularly the UASB reactors, in an outstanding position.

Nowadays, it can be said that the high-rate anaerobic reactors used for treatment of domestic sewage are a consolidated technology in some warm-climate countries, especially in Brazil, Colombia and India, with several treatment systems operating in full scale (population equivalents from a few thousand up to around one million inhabitants). In Brazil, practically all the wastewater treatment feasibility studies include anaerobic reactors as one of the main options. Undoubtedly, a great contribution to the consolidation and dissemination of the anaerobic technology for the treatment of domestic sewage came from the Brazilian National Research Programme on Basic Sanitation, PROSAB.

7.1.2 Main limitations

In spite of their great advantages, anaerobic reactors hardly produce effluents that comply with usual discharge standards established by environmental agencies. Therefore, the effluents from anaerobic reactors usually require a post-treatment step as a means to adapt the treated effluent to the requirements of the environmental legislation and protect the receiving water bodies.

The main role of the post-treatment is to complete the removal of organic matter, as well as to remove constituents little affected by the anaerobic treatment, such as nutrients (N and P) and pathogenic organisms (viruses, bacteria, protozoans and helminths).

(a)   Limitations regarding organic matter

Limitations imposed by environmental agencies for BOD are usually expressed in terms of effluent discharge standards and minimum removal efficiencies. These constraints are probably the cause that has mostly limited the use of anaerobic systems (without post-treatment) for sewage treatment (see typical values in Table 7.1).

In view of the limitations imposed by environmental legislation for the effluent BOD concentration, or also when the receiving body has limited capacity for assimilating the effluent from the treatment plant (which is frequently the case), it is usually necessary to use aerobic treatment to supplement the anaerobic stage. However, there are situations in which the combination of different anaerobic processes can meet less restrictive requirements regarding efficiency and concentration of the final effluent (e.g. 80% and 60 mgBOD/L, respectively). This is the case for systems consisting of a septic tank followed by an anaerobic filter (usually feasible for small populations, generally fewer than 1,000 inhabitants) or for a UASB reactor followed by an anaerobic filter. Obviously, the application of these combined anaerobic systems is conditioned to an appropriate dilution capacity of the receiving body.

Table 7.1. Usual effluent BOD and removal efficiencies in anaerobic systems

Anaerobic system	Effluent BOD (mg/L)	BOD removal efficiency (%)
Anaerobic pond	70 to 160	40 to 70
UASB reactor	60 to 120	55 to 75
Septic tank	80 to 150	35 to 60
Imhoff tank	80 to 150	35 to 60
Septic tank followed by anaerobic filter	40 to 60	75 to 85

Source: Chernicharo *et al.* (2001c)

In this sense, in situations in which the receiving body presents a good dilution capacity, the adoption of less restrictive discharge standards could enable the construction of simpler and more economical treatment plants in several small cities by means of a more intensive use of anaerobic reactors, particularly UASB reactors. At a later stage, if it becomes necessary to produce a better quality effluent, a complementary treatment unit can be built after some years. The high costs of sophisticated treatment systems, designed exclusively to meet BOD discharge standards, make their construction at a single stage unfeasible for most cities located in developing countries. On the other hand, the construction in stages could be decisive, in that systems consisting of a UASB reactor and a post-treatment unit become the most feasible ones regarding technical and economical criteria.

(b)   Limitations regarding nitrogen and phosphorus

The discharge of nutrients into surface water bodies may cause increased algal biomass as a result of the eutrophication process. It is known that 1.0 kg of phosphorus can result in the reconstruction of 111 kg of biomass, which corresponds to approximately 138 kg of chemical oxygen demand in the receiving body. Similarly, the discharge of 1.0 kg of nitrogen can result in the reconstruction of approximately 20 kg of chemical oxygen demand under the form of dead algae. The problem can be even worsened due to the decreased oxygen levels, by means of the nitrification processes, when at least 4.0 kg of dissolved oxygen are consumed for each kilogram of ammonia discharged into the receiving body.

In cases in which nutrient removal is required to meet the quality standards of the receiving water body, the use of anaerobic processes preceding a complementary aerobic treatment for biological nutrient removal should be analysed very carefully, once anaerobic systems present good biodegradable organic matter removal, but practically no N and P removal efficiency. This certainly causes an adverse effect on biological treatment systems aiming at good nutrient removal, because the effluent from the anaerobic reactor will have N/COD and P/COD ratios much higher than the values desired for good performance of biological nutrient removal processes (Alem Sobrinho and Jordão, 2001).

When the purpose of the treatment plant is also good nitrogen removal, the anaerobic reactor should be used to treat initially only a part of the influent raw sewage (possibly no more than 50 to 70%), and the remaining part (50 to 30%) should be directed to the complementary biological treatment, aiming at nitrification and denitrification, so that there is enough organic matter for the denitrification step. In this case, the great advantage of the use of the anaerobic reactor is that it can receive and stabilise the sludge generated in the complementary treatment, eliminating the need for an anaerobic sludge digester.

On the other hand, when the purpose is the biological phosphorus removal, the use of an anaerobic reactor is not advisable for two main reasons: (i) the effluent from the anaerobic reactor presents a P/COD ratio higher than that of the raw sewage, which harms the performance of the biological phosphorus removal system; and (ii) if the phosphorus-rich sludge generated in the biological phosphorus removal treatment is directed to the anaerobic reactor for stabilisation, the phosphorus incorporated to this sludge will be released under anaerobic conditions and leave with the effluent from the anaerobic reactor. This fact makes efficient phosphorus removal unfeasible in a treatment plant with an anaerobic reactor followed by complementary treatment with biological phosphorus removal.

According to Alem Sobrinho and Jordão (2001), phosphorus removal in treatment plants using an anaerobic reactor will only be effective if chemical products are used for P precipitation (iron or aluminium salts). In this case, the anaerobic reactor has the advantage of stabilising the sludge generated in the complementary biological aerobic treatment.

(c)   Limitations regarding microbiological indicators

Regarding the microbiological indicators, low faecal coliform removal efficiencies have been reported in anaerobic reactors, usually amounting to around only 1 log-unit. Regarding other types of microorganisms, such as viruses and protozoans (mainly *Giardia* and *Cryptosporidium*), there are few references covering their reduction or elimination in anaerobic reactors. The removal of helminth eggs in anaerobic reactors, particularly in UASB reactors, has been reported as amounting to 60 to 90%, being therefore insufficient to produce effluents that may be used in irrigation. However, it should be mentioned that these limitations are not exclusive of anaerobic reactors, but are a characteristic of most compact wastewater treatment systems.

As the risk of human contamination by ingestion or contact with water containing pathogenic organisms is high, many times it may be necessary to disinfect the effluents. This fact becomes even more serious due to the poor sanitary conditions in developing countries. On the other hand, the low investments in health and sanitation make the population of these countries bearers of several diseases that can be transmitted by faeces and, consequently, by the sewage generated by this population.

However, although the domestic sewage is an unquestionable source of contamination by pathogenic organisms, it is worth mentioning that the agents used in the disinfection processes can also cause harm to human health and the aquatic environment. It is then concluded that the decision whether or not to disinfect the sewage should be taken from a careful evaluation, based on the specific characteristics of each situation. In other words, there are no universal guidelines ruling sewage disinfection requirements. The decision on the need to disinfect the sewage of a certain locality involves (USEPA, 1986):

- an investigation on the uses of the water downstream the discharge point, and on the public health risks associated with that water

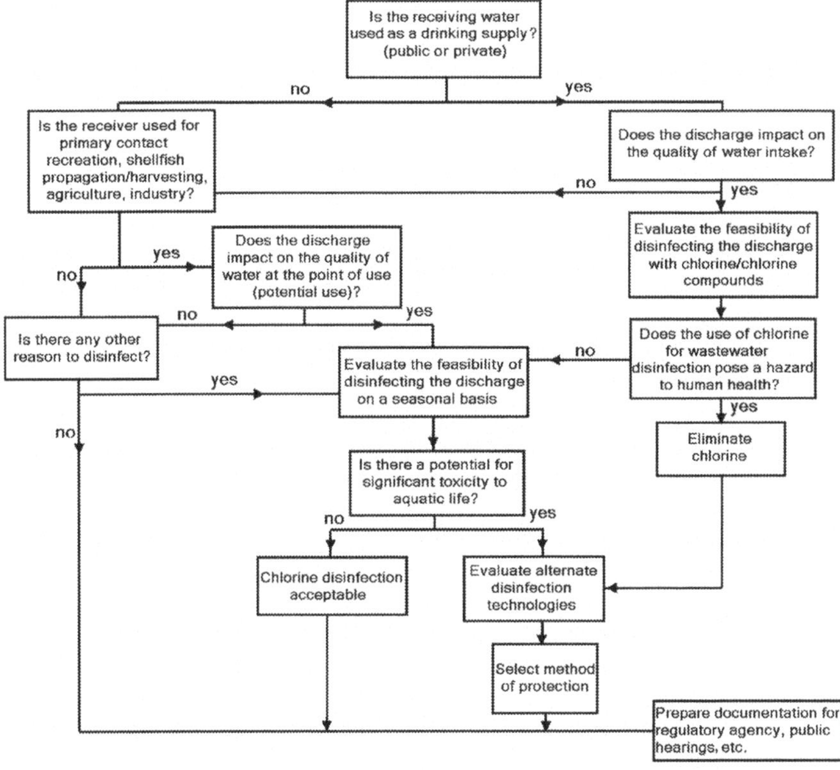

Figure 7.1.   Flowchart for local evaluation of the need for and requirements of sewage disinfection (adapted from USEPA, 1986)

- an evaluation of the alternatives available for control of the sewage contaminated by pathogens
- an evaluation of the environmental impacts the control measures may cause

Figure 7.1 presents a flowsheet that can aid the decision making on the implementation need and requirements of a sewage disinfection system, taking into account the public health risks involved and the possibility of either reducing or eliminating these risks. Once the risks involved are identified, the environmental aspects start to determine the applicability of the control alternative.

7.1.3 Advantages of the combined (anaerobic/aerobic) treatment

In comparison with a conventional wastewater treatment plant consisting of a primary sedimentation tank followed by aerobic biological treatment (activated sludge, trickling filter, submerged aerated biofilter, or biodisc), with the primary and secondary sludge passing through sludge thickeners and anaerobic digesters prior to dewatering, a treatment consisting of a UASB reactor followed by aerobic biological treatment (with the secondary sludge directed to thickening and digestion in the UASB reactor itself and then straight to dewatering) can present the following advantages (Alem Sobrinho and Jordão, 2001):

- The primary sedimentation tanks, sludge thickeners and anaerobic digesters, as well as all their equipment, can be replaced with UASB reactors, which do not require the use of equipment. In this configuration, besides their main sewage treatment function, the UASB reactors also accomplish the aerobic sludge thickening and digestion functions, requiring no additional volume.
- Power consumption for aeration in activated sludge systems preceded by UASB reactors will be substantially lower compared to conventional activated sludge systems, and especially extended aeration systems.
- Thanks to the lower sludge production in anaerobic systems and to their better dewaterability, sludge volumes to be disposed of from anaerobic–aerobic systems will be much lower than those from aerobic systems alone.
- The construction cost of a treatment plant with a UASB reactor followed by aerobic biological treatment should be no more than 80% of the cost of a conventional treatment plant. In addition, due to the simplicity, smaller sludge production, and lower power consumption of the combined anaerobic–aerobic system, the operational costs also represent an even greater advantage.

7.2 MAIN ALTERNATIVES FOR THE POST-TREATMENT OF EFFLUENTS FROM ANAEROBIC REACTORS

7.2.1 Preliminaries

Taking into consideration the intrinsic limitations associated with the anaerobic systems and the need to develop technologies that are more appropriate to the reality of developing countries, it is important to include a post-treatment stage for the effluents generated in anaerobic reactors. This stage has the purpose of polishing not only the microbiological quality of the effluents, in view of the public health risks and limitations imposed on the use of treated effluents in agriculture, but also the quality in terms of organic matter and nutrients, in view of the environmental damages caused by the discharges of the remaining loads of these components into the receiving bodies.

Considering that the treatment line consisting of *anaerobic reactors+post-treatment units* is an important alternative for developing countries, the main progresses achieved on this subject by the Brazilian National Research Programme on Basic Sanitation, PROSAB (Chernicharo *et al.*, 2001c) are presented in this chapter. The main aspects of the most important post-treatment alternatives being applied in Brazil are herein discussed.

7.2.2 Anaerobic filter

7.2.2.1 Preliminary considerations

The main innovative purpose of the research was to evaluate the applicability of an anaerobic process (anaerobic filter) used for the polishing of domestic sewage, whose previous treatment stage is also performed by another anaerobic process (UASB reactor). This association of anaerobic processes contributes greatly to the reduction of power and operational costs of the treatment plant.

Until recently, the anaerobic filters were limited to small populations, usually treating effluents from septic tanks. Nowadays, anaerobic filters after UASB reactors are being used to produce a final effluent with BOD lower than 60 mg/L, even in cities with population larger than 50,000 inhabitants. The complementary organic matter removal achieved in the second anaerobic reactor (anaerobic filter) occurs by:

- the retention of solids in the anaerobic filter, reflecting on the removal of particulate organic matter. In this case, physical removal mechanisms prevail through the combined effects of coarse filtration in the packing medium and sedimentation along the column
- the formation of biofilm on the packing medium and removal of the remaining soluble organic matter. In this case, the formation of biofilm and the removal of carbonaceous matter by biochemical means depend on the amount of organic matter present in the effluent from the UASB reactor.

7.2.2.2 Typical configuration

Wastewater treatment plants using UASB reactors followed by anaerobic filters represent a very simple flowsheet (Figure 7.2). Besides the preliminary treatment

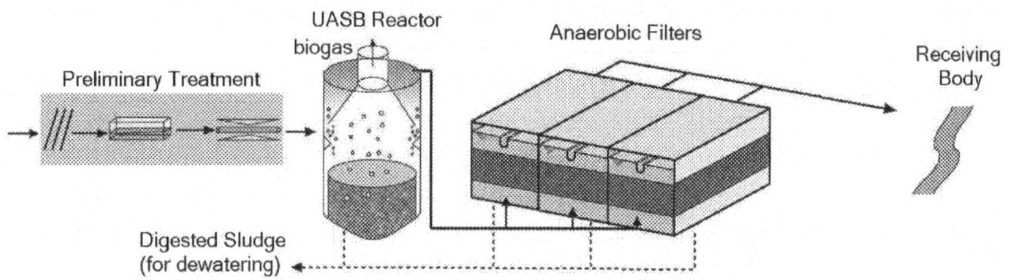

Figure 7.2.   Typical configuration of a treatment plant with a UASB reactor and an anaerobic filter

units (screen and grit chamber), the flowsheet comprises basically the two sequential anaerobic treatment units (UASB reactor and anaerobic filter) and the dewatering unit. This is because the sludge produced in the anaerobic units are already thickened and stabilised. Sludge drying beds have been frequently used for sludge dewatering in small plants. UASB reactor+anaerobic filter facilities have already been installed in some locations in Brazil, as shown in Figures 5.3 and 5.4.

7.2.2.3 Design criteria

A deep discussion on the main design criteria and parameters for anaerobic filters is presented in Chapter 5. These criteria were obtained from pilot-scale research and from operational results from full-scale plants.

7.2.3 Polishing ponds

7.2.3.1 Preliminary considerations

Facultative ponds are largely used for post-treatment of effluents from anaerobic ponds. These systems have the advantage of removing at a higher efficiency the pathogenic organisms present in the sewage, but their main disadvantage is the high concentration of algae in the final effluent, which leads to serious restrictions by some environmental agencies.

When an efficient anaerobic pre-treatment is applied prior to the sewage discharge into a pond, the concentrations of organic matter and suspended solids are largely reduced, and consequently only a complementary removal of these two constituents will be required, needing much lower hydraulic detention times. In these conditions, the limiting factor that determines the minimum detention time (and, therefore, the volume and the area of a pond system) will usually be the removal of pathogenic organisms, and not the stabilisation of the organic matter. For this reason, the nomenclature **polishing pond** has been adopted to name those ponds intended for the post-treatment of effluents from efficient anaerobic systems, thus distinguishing them from the **stabilisation pond,** which treats raw sewage (Cavalcanti *et al.*, 2001).

The UASB reactor+polishing pond configuration is a very interesting alternative from the technical–economical–environmental point of view, mainly when there are area limitations for the construction of only stabilisation ponds. In addition, the problems related to odours from anaerobic ponds can be avoided in plants utilising a UASB reactor and polishing pond, since the anaerobic reactor can be installed with odour control. This alternative is even more attractive when the effluent from the pond can be used for agricultural purposes, since the polishing ponds aim mainly at the removal of pathogenic organisms. Because of its advantages, the post-treatment of effluents from anaerobic reactors through ponds has been common in developing countries.

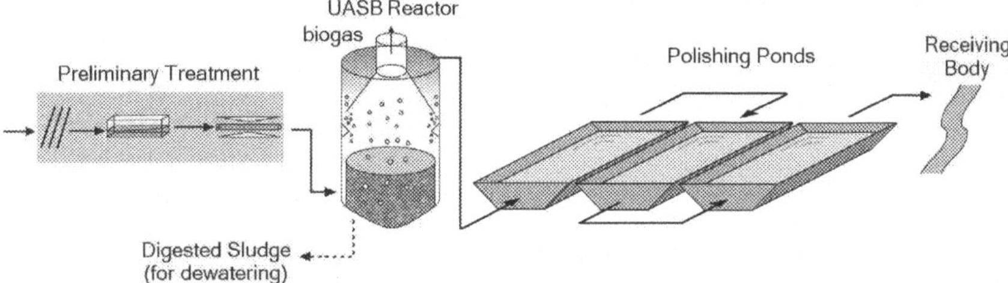

Figure 7.3. Typical configuration of a treatment plant with a UASB reactor and polishing ponds

Figure 7.4. View of a UASB reactor followed by four polishing ponds in series (250 inhabitants, Arrudas Experimental WWTP, UFMG/COPASA, Brazil)

7.2.3.2 Typical configuration

Wastewater treatment plants using UASB reactors followed by polishing ponds also have a very simplified flowsheet (Figure 7.3). Besides the preliminary treatment units (screen and grit chamber), the flowsheet comprises the anaerobic treatment unit, the polishing pond (either a single baffled pond or ponds in series) and the dewatering unit for the sludge produced in the UASB reactor. The same considerations made for the UASB reactor+anaerobic filter system are valid here in relation to the characteristics of the anaerobic sludge, which is already thickened and stabilised. Thus, dewatering units using drying beds are also usual in smaller plants. Figure 7.4 illustrates a research unit implemented by the Federal University of Minas Gerais, Brazil.

7.2.3.3 Design criteria

These criteria were obtained from research at pilot and demonstration scales and from operational results from full-scale plants.

7.2.4 Land disposal

7.2.4.1 Preliminary considerations

Land disposal of sewage is an ancient practice, in which filtration and the action of microorganisms take place. The microorganisms have the capacity to convert the organic matter into simpler compounds. A treated effluent and a revitalised soil are obtained as a final result of this process, since the compounds generated by the microorganisms can be beneficial for the growth of plants and vegetables.

The current section covers only *overland flow* systems as a means of post-treatment of effluents from UASB reactors. A detailed description, the typical configuration and the main design criteria for the other systems can be found in Coraucci Filho *et al.* (2001).

Sewage treatment by the *overland flow* method is the one that presents the least dependence on the types of soil. In this method, the vegetation, associated with the top soil layer, acts as a filter, removing the nutrients and providing conditions for the retention and transformation of the organic matter contained in the sewage. Besides that, it protects the soil against erosion and creates a support layer on which the microorganisms settle. The main mechanisms through which organic matter and solids are removed are biological oxidation, sedimentation and filtration. The main characteristic that differentiates this method from the others is the fact that the effluent flows downward on a slightly inclined vegetated ramp and the remaining water (effluent), which is neither absorbed nor evaporated, is collected downstream and directed for disposal. For more permeable soils, the process is similar to that of irrigation, but with the generation of effluent.

In comparison with other land disposal methods, *overland flow* presents the following characteristics as its main advantages (Coraucci Filho *et al.*, 2001):

- it is appropriate for the treatment of sewage from rural communities and from seasonal industries that generate organic wastewater
- it provides an advanced secondary treatment, with a relatively simple, cheap operation
- the vegetable covering can be reused or commercially used
- it presents the minimum restriction regarding the characteristics of the land, requiring only relatively impermeable soil for its installation and an adequate slope

The disadvantages are:

- the method is limited by the climate, culture tolerance in relation to water and slope of the land
- the application may be limited during wet weather
- the loading rates may be restricted by the growth pattern of the culture
- flat or very steep land is not suitable for this type of treatment

Therefore, the method consists in applying the liquid in the highest part of the ramp. The effluent then drains all over the slope by gravity, where part of it is lost by evapotranspiration and the remaining part is collected on the base of the ramp. Percolation can be insignificant because this system is initially conceived for low-permeability soils. In spite of that, its use has been also reported for soils with medium permeability and impermeable underground (USEPA, 1981). Sewage application is intermittent and the following types of feeding can be adopted: (i) high-pressure sprinklers; (ii) low-pressure sprinklers; and (iii) distribution piping or channels with spaced openings.

Organic matter removal. The effluent produced by overland flow treatment systems usually presents low BOD concentrations. BOD is removed by the biofilm that grows on the surface of the soil and plants. The biofilm can eventually become very thick due to excessive growth. The bacterial cells close to the surface of the soil and plants die due to the lack of oxygen. Different from other attached growth treatment systems, the dead mass of biological solids is not significantly removed from the system, being eventually degraded as time goes by. The complete development of the biofilm after the system start-up may take some time, even 1 year in some cases (WPCF, 1990).

The experiences using the *overland flow* method for the post-treatment of anaerobic effluents have indicated BOD and COD removal efficiencies in the ramps ranging from 48 to 52%, depending on the applied loading rates (Chernicharo *et al.*, 2001a). The overall efficiency of the *anaerobic reactor+ overland flow system* usually amounts from 80 to 90%.

Suspended solids removal. The removal of suspended solids is very efficient in overland flow systems, due to the reduced flow velocities over the ground (between 0.3 and 3 cm/s). The solid material removed works as a substrate for the biofilm, being virtually degraded.

Nitrogen removal. The mechanisms responsible for the removal of nitrogen in overland flow systems include absorption by plants, nitrification/denitrification and ammonia stripping. The plants are capable of removing between 20 and 30% of the total N (e.g. Martel *et al.*, 1980). The removal rate by plants depends on the vegetation culture selected, on the depth and distribution of the roots, on the N loading rate, on the movement of water in the soil and other factors. In general, a type of grass that takes time to develop and presents high nitrogen absorption rates is chosen. It is recommended that the vegetation is periodically harvested, to obtain higher efficiencies.

The losses by ammonia volatilisation are very variable and present a close relation with the evaporation rate and the sewage loading technique. The application of effluents by means of high-pressure sprinklers results in the loss of approximately 7 to 11% of nitrogen in the form of ammonia, while ammonia stripping during the flow of the effluent on the soil is usually lower than 5% (Khalid *et al.*, 1978).

The nitrification process is mainly affected by the amount of oxygen available, the loading rate, the pH and temperature. In mild climates, the limiting factors are the amount of available oxygen and the loading rate. The ratio between the wet and dry periods controls the availability of oxygen in the medium and the time necessary for nitrification. The loading rate is inversely proportional to the ammonia removal, that is, the higher the loading rate, the lower the ammonia removal efficiency. The denitrification process is affected by the degree of treatment of the wastewater applied; once that happens, the higher the concentration of influent BOD to the treatment system, the larger the probability of development of anaerobic conditions and the presence of carbonaceous matter sufficient for denitrification. The BOD_5:N ratio should be approximately 3:1, to favour better removal efficiencies.

The experiences with the use of the *overland flow* process for the post-treatment of anaerobic effluents in Brazil have indicated nitrogen removal efficiencies ranging from 75 to 90%, depending on the temperature, sewage loading rates, and feeding and resting times.

Phosphorus removal. Phosphorus removal in overland flow systems occurs by sedimentation and adsorption in the soil and plants. Removal rates vary between 20 and 60%, although values in the range of 84 to 89% have already been reported (Lee *et al.*, 1976; Martel *et al.*, 1980). Approximately 10% of the phosphorus, corresponding to the insoluble part, is removed in the previous treatment system (in this case, the anaerobic reactor). Except for the component that is incorporated to the biomass, the additional phosphorus removal is minimum in the conventional biological treatment systems, since most of the phosphorus present after the primary treatment is in soluble form. Phosphorus removal in overland flow systems is not usually high, due to the limited contact existing between water and soil, hindering the adsorption process.

Pathogenic organism removal. The survival of pathogenic bacteria in the soil is subject to several factors, including the antagonism of the microflora, moisture content, water retention capacity, organic matter concentration, pH, solar radiation and temperature (Feachem *et al.*, 1983). In overland flow systems, the main microorganism removal mechanisms include: sedimentation; filtration through the biofilm formed on the stems of plants and on the upper layer of the soil; adsorption by soil particles; predation; solar irradiation and desiccation.

In general, and according to experimental results obtained in the past, it can be said that overland flow systems are not efficient regarding the removal of microbial indicators, such as faecal (thermotolerant) coliforms (WPCF, 1990). Peters and Lee (1978) observed a reduction of just one logarithmic unit (or a 90% reduction) in the faecal coliform levels after the application of raw wastewater to an overland flow system. Chernicharo *et al.* (2001a) obtained slightly better results in experiments conducted in a *UASB+overland flow system* treating domestic sewage, in which the removals of faecal coliforms were one log-unit for the UASB reactor and one to two log-units for the overland flow system, resulting in a final effluent with concentrations in the range of 10^4 to 10^5 MPN/100 mL.

The existing knowledge on virus survival in the soil, which is not very comprehensive yet, suggests that the protein nature of these microorganisms favours their adsorption onto the surface of the soil particles (mainly if the soil is of clayey nature), where they are protected from adverse environmental conditions (e.g. Goyal and Gerba, 1979). Schaub *et al.* (1978) observed enteric virus removal rates of up to 85% in overland flow systems.

Helminth eggs remain viable in the soil during long periods, although this varies from species to species. For instance, it is known that *A. lumbricoides* and *T. saginata* eggs can survive in the soil for periods longer than those necessary for plant growth. Vegetable cultures irrigated with wastewater from regions where ascariasis and teniasis are endemic are a potential disease transmission risk (WHO, 1985). Stien and Schwartzbrod (1990) concluded from an experimental study in laboratory scale that the survival time of *Ascaris* eggs in the soil decreases quickly after 20 days from the date of contamination by artificial wastewater. The egg elimination process in the soil depends essentially on two factors: exposure to sunlight and type of soil. Eggs were not found in the vegetable samples after 10 days from wastewater application. The survival time of the eggs in the roots depends on the type of vegetable culture but, in general, it decreases quickly after 45 days from the contamination. Chernicharo *et al.* (2001a) observed no helminth eggs in the final effluent of an overland flow system fed with domestic sewage previously treated in a UASB reactor.

The main characteristics and results of experiments with overland flow systems used for the post-treatment of effluents from anaerobic reactors in Brazil are presented in Table 7.2 (Coraucci Filho *et al.*, 2001).

7.2.4.2 Typical configuration

The typical configuration of a wastewater treatment plant consisting of a UASB reactor and post-treatment by overland flow has a very simple flowsheet (Figure 7.5). Besides the preliminary treatment units (screens and grit chambers), the flowsheet comprises the anaerobic treatment unit, the land treatment system and the dewatering unit for the sludge produced in the UASB reactor. The same considerations made for the systems previously discussed, regarding the characteristics of the anaerobic sludge that is already thickened and stabilised, are also valid here. Dewatering units using drying beds can be used in small-sized plants.

7.2.4.3 Design criteria

The main criteria for the design of overland flow systems applied to the post-treatment of effluents from anaerobic reactors are as follows (adapted from USEPA, 1981; WPCF, 1990 and Coraucci Filho *et al.*, 2001):

Table 7.2. Characteristics and results of experiments with post-treatment systems by overland flow

Parameter	Experiment 1	Experiment 2	Experiment 3	Experiment 4
Type of pre-treatment system	Anaerobic filter	Anaerobic filter	UASB reactor	UASB reactor
Width of the slope (m)	4.2	4.2	3.0	3.0
Length of the slope (m)	35	35	25	25
Gradient of the slope (%)	3.5	3.5	4	4
Hydraulic loading rate (m^3/hour·m)	0.10 and 0.20	0.30 and 0.40	0.20 to 0.60	0.48[a]
Feeding period (hour/d)	8	8	8	8
Feeding frequency (d/week)	5	5	5	5
Vegetation cover	*Tifton 85*	*Tifton 85*	*B. humidicola*	*Tifton 85*
Average characteristics of the final effluent				
BOD (mg/L)	30	60	48 to 62	60
COD (mg/L)	116	–	98 to 119	–
TSS (mg/L)	40	–	17 to 57	–
TKN (mg/L)	13	–	–	14 to 18
P (mg/L)	0.5	–	–	–
E. coli (MPN/100 mL)	–	–	–	10^4 to 10^5
Helminth eggs (egg/L)	–	–	0.2	0

(a) Average rate (variable flow over the day, due to the transient hydraulic feeding system to the slopes)
Source: Adapted from Coraucci Filho *et al.* (2001)

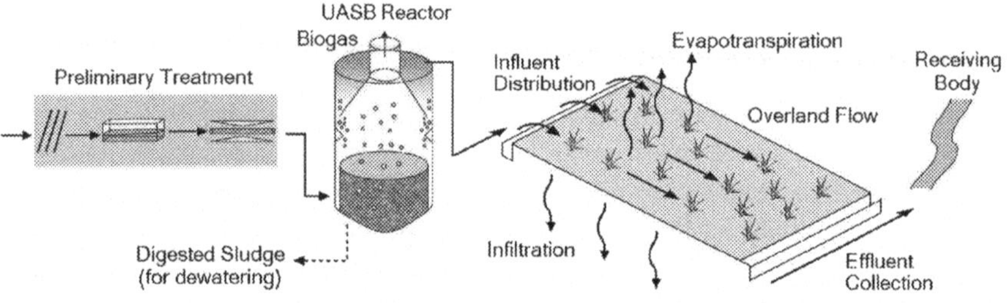

Figure 7.5.   Typical configuration of a treatment plant with a UASB reactor and overland flow system

Length of the slope. The length is the longitudinal dimension of the physical surface of the soil, defined by the flowing direction of the effluent. For the low-pressure wastewater application technique, the length of the slope ranges from **30 to 45 m**. Lengths between 45 and 60 m are used for the high-pressure distribution systems.

Ground slope. A ground slope between 1 and 12% is recommended, with an optimal interval between **2 and 8%**. A slope lower than 1% is not recommended, due to the possible formation of pools with sewage and the consequent proliferation of flies. Very high slopes cause the decrease of the flow time and the treatment efficiency, besides favouring the development of erosive processes.

Classification of the soil. The overland flow system was initially developed for soils with low permeability, lower than 15 mm/hour. In spite of that, the system can be used in locations with moderate permeability (**15 to 50 mm/hour**). This is because the void spaces of the soil can be filled with influent solids (clogging) and vegetable growth over time. The permeability can also be changed by soil compaction during the construction of the system.

Operation cycle. The operation is intermittent, with a *feeding period* between **8 and 12 hours/d**, followed by a *dry period* ranging from **16 to 24 hours/d**. Operational cycles with **4 days feeding and 2 days resting (dry)** avoid the propagation of insects.

Hydraulic loading rate. The loading rate is considered the main parameter for the design of the system, defined as the volume applied to the treatment module divided by the loading period in hours. There is a tendency to standardise this parameter, expressing it in terms of unit-width of the module, in $m^3/hour \cdot m$ (Paganini, 1997; Coraucci Filho *et al.*, 2001). This parameter is dependent on the effluent discharge regime, on the sewage pre-treatment level, on the depth and slope of the ground, as well as on the climate. For the post-treatment of anaerobic effluents, the use of loading rates between **0.2 and 0.4 $m^3/hour \cdot m$** of width of the slope has been usual.

7.2.4.4 Construction aspects

The following main aspects in relation to the construction of overland flow systems should be taken into consideration (USEPA, 1981; WPCF, 1990 and Coraucci Filho *et al.*, 2001):

Storage. It is necessary to build a storage tank sufficient to store the effluent on the days when there is no application. The liquid should be stirred during this period.

Distribution of the sewage. The uniform distribution of the wastewater on the whole width of the ramp is a critical factor in the performance of the system. Its application by either low- or high-pressure sprinklers or by perforated tubes should be started from the top of each slope. The effluent can be distributed by three different techniques (see also Table 7.3):

- *piping with spaced openings:* piping similar to that used for irrigation. The influent is applied under low pressure (2 to 5 N/cm^2). An adjustment should

Table 7.3. Distribution methods: advantages and limitations

Method	Advantage	Limitation
Piping with adjustable openings	• Easy cleaning • Low power consumption • Little generation of aerosols • Smaller safety areas • Easier water balance control	• Possibility of sedimentation inside the tubes • Difficult uniform distribution • Possibility of erosion • Blocking of the orifices
Cut or perforated piping	• Low power consumption • Little generation of aerosols • Smaller safety areas	• Difficulty to ensure uniform distribution • Possibility of erosion • Difficulty to control the water balance • Blocking of the orifices
Bubbling orifice	• Low power consumption • Little generation of aerosols • Smaller safety areas • Less susceptibility to sedimentation	• Difficulty to achieve uniform distribution • Possibility of erosion • Difficulty in maintenance when blocked
Distribution channels	• Low power consumption • Little generation of aerosols • Smaller safety areas • Easy operation	• High initial construction cost • Possibility of erosion • Formation of preferential routes
Low-pressure sprinklers	• More uniform sewage distribution • Low power consumption • Production of less aerosols than high-pressure sprinklers	• Possibility of orifice obstruction by large particles • Generation of aerosols
High-pressure sprinklers	• More uniform sewage distribution • Fewer maintenance requirements	• High power consumption • Larger generation of aerosols • Larger safety areas

Source: Adapted from Araújo (1998)

be made to obtain a uniform distribution. This type of distribution is not recommended for influents with high concentration of suspended solids due to the potential deposition of solids close to the discharge point

- *low-pressure sprinklers:* used with pressures between 5 and 15 N/cm^2. In this type of distribution, the solids can cause the blockage of the sprinkler openings
- *high-pressure sprinklers:* used with pressures between 35 and 60 N/cm^2. This type of distribution covers larger areas than those previously mentioned. As the effluent can reach longer distances, the construction of longer slopes is recommended, to have an appropriate treatment. However, care

should be taken in the use of this type of sprinkler in the case of domestic sewage, in view of the contamination risks by aerosols

Selection of the vegetation. The covering vegetation is essential to the good performance of the system. Perennial and water resistant grasses are those that adapt better to overland flow systems. Their main functions are: protection against erosion, redistribution of the flow (which avoids short circuits), support for microorganisms and removal of nutrients.

Monitoring. The flow, the applied rates, the period and frequency of sewage loading, and the quality of the influent and effluent should be constantly monitored. If there is significant infiltration into the soil, the groundwater shall also be monitored.

Example 7.1

Design an overland flow system acting as post-treatment of the effluent from a UASB reactor, with the following design elements being known:
Data:

- Population: $P = 20,000$ inhabitants
- Average influent flow: $Q_{av} = 3,000$ m^3/d (125 m^3/hour)
- Average influent BOD (S_o) = 350 mg/L

The anaerobic reactor was designed in Example 5.2.

Solution:

(a) Calculation of the required area

Design parameters (see Section 7.2.4.3):

- Loading rate: $q_L = 0.35$ m^3/hour·m
- Length of the slope: $Z = 35$ m
- Feeding periods (feeding hours per day in each slope): $L_p = 8$ hours/d
- Feeding frequency (loading days per week): $f = 5$ d/week

Net area required:

$$A = \frac{Q_{av} \times Z}{q_L \times L_p} \times \left(\frac{7}{f}\right)$$

$$= \frac{(3,000 \text{ m}^3/\text{d}) \times (35 \text{ m})}{(0.35 \text{ m}^3/\text{m·hour}) \times (8 \text{ hours/d})} \times \left(\frac{7 \text{ d/week}}{5 \text{ d/week}}\right) = 52,500 \text{ m}^2$$

Total area (assuming a 20% increment for urbanisation, roads, laboratory, interconnections, etc):

Total area = $1.2 \times 52,500$ m^2 = $63,000$ m^2 = (6.3 ha)

Per capita land requirement = (63,000 m^2)/(20,000 inhabitants) = 3.2 m^2/inhabitant

---

Example 7.1 (Continued)

(b)   Dimensions of each slope

Number of slopes (initial trial value; this value can be revised, to allow more favourable dimensions and a better adjustment among units in terms of load, daily rest and weekly rest): n = 25

Area of each slope: $A_u = A/n = (52{,}500\ m^2)/25 = 2100\ m^2$

Length of each slope: Z = 35.00 m (previously defined, design parameter)

Width of each slope: $W = A_u/Z = (2100\ m^2)/(35.00\ m) = 60.00\ m$

Gradient of the slopes: s = 4% (design parameter, see Section 7.2.4.3)

Level difference between the upper and the lower parts of each ramp: $H = (Z{\cdot}s/100) = 35.00\ m \times 4/100 = 1.40\ m$

(c)   Operational regime of the slopes

Weekly cycle:

- Number of slopes in rest: $n_r = n{\cdot}(1 - f/7) = 25 \times (1 - 5/7) = 7$

Daily cycle:

- Number of slopes in operation: $n_{op} = n - n_r = 25 - 7 = 18$
- Number of slopes in loading (at each instant): $n_{load} = n_{op}{\cdot}L_p/24 = 18 \times 8/24 = 6$
- Number of slopes in resting (at each instant): $n_r = n_{op} - n_{load} = 18 - 6 = 12$

(d)   Concentration of effluent BOD

Effluent concentration of the UASB reactor (assuming E_{UASB} efficiency = 75%):
$BOD_{efflUASB} = 350\ mg/L{\cdot}(1-75/100) = 88\ mg/L$ (see Example 5.2)

Effluent concentration of the overland flow (assuming E = 50%):
$BOD_{effl} = 88\ mg/L{\cdot}(1-50/100) = 44\ mg/L$

Overall efficiency of the system:
$E = (350 - 44)/350 = 0.87 = 87\%$

7.2.5 Trickling filter

7.2.5.1 Preliminary considerations

A trickling filter consists basically of a tank filled with a highly permeable material, onto which wastewater is loaded in the form of drops or jets. Wastewater percolates towards the bottom drains, allowing bacterial growth on the surface of the packing material, in the form of a fixed film (biofilm). Wastewater passes over the biofilm, allowing a contact between the microorganisms and the organic matter.

Although the trickling filters (TF) are wastewater treatment systems with great potential and numerous advantages, mainly because of their simplicity and low operational costs, few units have been implemented so far with the purpose of performing the post-treatment of effluents from anaerobic reactors.

The main and innovative purpose of the researches carried out in the past years was to evaluate the applicability and behaviour of the trickling filters, when used for polishing of effluents from anaerobic reactors, particularly UASB reactors. This association (UASB reactor+TF) may contribute significantly to the reduction of the power and operational costs of the treatment plant.

7.2.5.2 Typical configuration

Wastewater treatment plants that use UASB reactors followed by trickling filters present a simple flowsheet (Figure 7.6). Basically, besides the preliminary treatment units (screens and grit chambers), the flowsheet comprises the sequential anaerobic and aerobic biological treatment units (UASB reactor, trickling filter and secondary sedimentation tank), as well as the dewatering unit. Notice that, in this configuration, the excess aerobic sludge removed from the secondary sedimentation tank is returned to the UASB reactor for thickening and anaerobic digestion. Therefore, with this flowsheet, primary sedimentation tanks and separate units for thickening and anaerobic digestion of the excess aerobic sludge are not required, different from the conventional treatment plants that use trickling filters.

The sludge wasted from UASB reactors is already thickened and stabilised, and can be sent directly for dewatering and final disposal. Drying beds have been frequently used for dewatering of the sludge in small-sized plants.

An innovative and compact configuration of this treatment system was developed by the Federal University of Minas Gerais (Brazil) for sewage treatment in

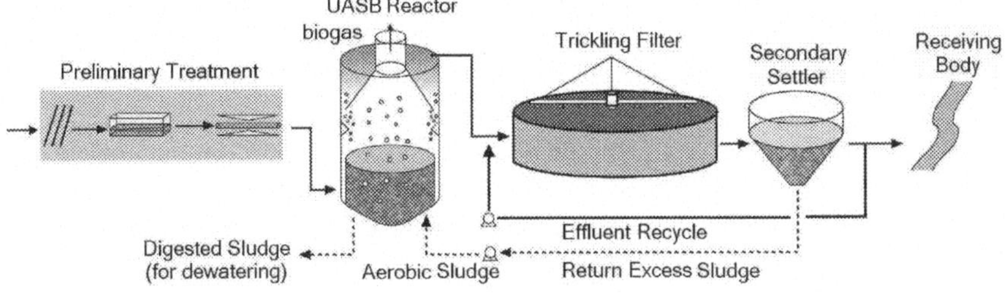

Figure 7.6.   Typical configuration of a treatment plant with UASB reactor and trickling filter

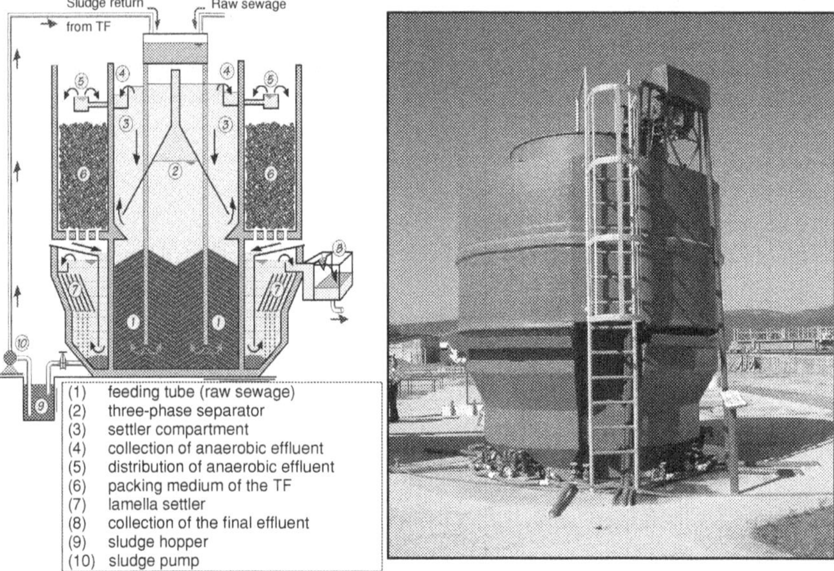

Figure 7.7.   Compact configuration of a UASB reactor and trickling filter system (module for 500 inhabitants, Arrudas Experimental WWTP, Brazil)

small communities. The compact system comprises the three main units (UASB reactor and TF reaction and settling compartments) in a single treatment module, as illustrated in Figure 7.7.

7.2.6 Submerged aerated biofilter

7.2.6.1 Preliminary considerations

A submerged aerated biofilter consists of a tank filled with porous material, through which sewage and air flow permanently. In almost all the existing processes, the porous medium is maintained totally submerged by the hydraulic flow. The biofilters are characterised as three-phase reactors consisting of:

- solid phase: consisting of the support medium and colonies of microorganisms present in the form of a biofilm
- liquid phase: consisting of the liquid in permanent flow through the porous medium
- gas phase: formed by artificial aeration and, in a reduced scale, by the gases derived from the biological activity

Several small wastewater treatment plants with UASB reactors followed by submerged aerated biofilters filled with granular material, without secondary sedimentation tanks, and with backwashing removal of sludge from the biofilter, are

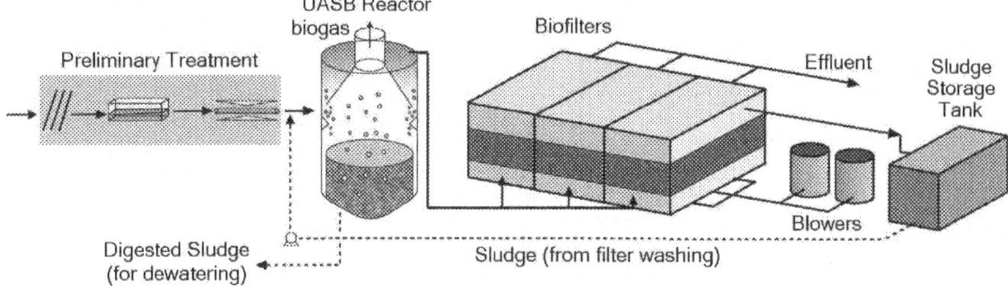

Figure 7.8.   Typical configuration of a treatment plant with UASB reactor and submerged aerated biofilters

already in operation in Brazil. Most of the plants have been designed for organic matter removal (effluent BOD < 30 mg/L), without nitrification.

7.2.6.2 Typical configuration

Sewage treatment plants that use UASB reactors followed by submerged aerated biofilters also present a simple flowsheet (Figure 7.8). Besides the preliminary treatment units (screens and grit chambers), the flowsheet comprises the sequential anaerobic and aerobic biological treatment units (UASB reactor and submerged aerated biofilter), as well as the aeration, sludge accumulation and dewatering units. Also in this configuration, the excess aerobic sludge removed from the biofilter is returned to the UASB reactor for thickening and anaerobic digestion. Therefore, with this flowsheet, primary sedimentation tanks and separate units for thickening and anaerobic digestion of the excess aerobic sludge are avoided, different from the conventional treatment plants that use submerged aerated biofilters.

The sludge wasted from the UASB reactor is already thickened and stabilised, and can be directly sent for dewatering and final disposal. Sludge drying beds have been frequently used in small-sized plants.

7.2.7 Activated sludge

7.2.7.1 Preliminary considerations

The essence of the *continuous flow* activated sludge process is the integration of the aeration tank (aerobic biological reactor), secondary sedimentation tank and sludge recirculation line. These three components are maintained in the alternative of activated sludge systems acting as post-treatment of effluents from anaerobic reactors.

The *intermittent flow* activated sludge system (sequencing batch reactors) can also be adopted as post-treatment, requiring, in this case, only the tanks that alternate in the functions of reaction and sedimentation.

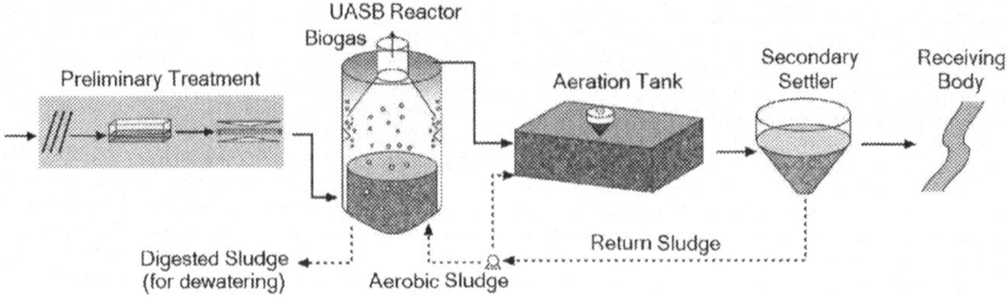

Figure 7.9.   Typical configuration of a treatment plant with UASB reactor and activated sludge system

7.2.7.2 *Typical configuration*

When the activated sludge system acts as post-treatment of anaerobic effluents, the anaerobic reactor is used instead of the primary sedimentation tank (which is an integral part of the conventional activated sludge system). The aerobic sludge is recirculated in the usual manner, that is, from the bottom of the secondary sedimentation tank to the entrance of the aerobic reactor (aeration tank).

The excess aerobic sludge generated in the activated sludge stage, not yet stabilised, is sent to the UASB reactor, where it undergoes thickening and digestion, together with the anaerobic sludge. As the return flow of the excess aerobic sludge is very low compared with the influent flow, there are no operational disturbances in the UASB reactor. The sludge treatment is largely simplified: there is no need for separate thickeners and digesters, and just the dewatering stage is necessary. The mixed sludge removed from the anaerobic reactor is digested, has solids concentrations similar to those from sludge thickeners and presents good dewaterability. Figure 7.9 presents the flowsheet of this configuration.

References

ABNT (1993). *Projeto, construção e operação de tanques sépticos.* NBR:7229, Rio de Janeiro,15 p. (in Portuguese).

Alem Sobrinho, P. and Jordão, E.P. (2001) Pós-tratamento de efluentes de reatores anaeróbios – uma análise crítica. In *Pós-tratamento de efluentes de reatores anaeróbios,* (C.A.L. Chernicharo coordenador), cap. 9 FINEP/PROSAB, Rio de Janeiro, Brasil, 544 p. (in Portuguese).

Anderson, G.K., Campos, C.M.M., Chernicharo C.A.L. and Smith, L.C. (1991) Evaluation of the inhibitory effects of lithium when used as tracer for anaerobic digesters. *Water Res.* **25**, 755–760.

Andreoli, C.V., Ferreira, A.C., Chernicharo, C.A.L. and Borges, E.S.M. (2003) Secagem e higienização de lodos com aproveitamento do biogás. In *Digestão anaeróbia de resíduos sólidos orgânicos e aproveitamento de biogás,* (Cassini S.T. coordenador), cap. 5 FINEP/PROSAB, Rio de Janeiro, Brasil, 196 p. (in Portuguese).

Araújo, G.C. (1998) Avaliação do pós-tratamento de efluentes de reatores UASB através de um sistema de aplicação superficial de esgotos no solo. Dissertação de mestrado, Universidade Federal de Minas Gerais, Brazil, 130 p. (in Portuguese).

Barredo Alcocer, M.D.S. (1988) The effect of pH on microbial interactions in the presence of propionic acid in anaerobic digestion. Ph.D. thesis, University of Newcastle upon Tyne, UK.

Barriga, L.H.M. (1997) El efecto de la humedad en la remocion del H_2S utilizando un filtro de compost. Dissertação de mestrado, Faculdad de Ingenieria, Universidad del Valle, Cali, Colômbia. (in Spanish).

Batstone, D.J., Keller, J., Angelidaki, I., Kalyuzhnyi, S.V., Pavlostathis, S.G., Rozzi, A., Sanders, W.T.M, Siegrist, H. and Vavilin, V.A. (2002) *Anaerobic Digestion Model No. 1 (ADM1).* IWA Scientific and Technical Report No. 13, IWA Publishing, London, 77 p.

Belli Filho, P., Costa, R.H.R., Gonçalves, R.F., Coraucci Filho, B. and Lisboa, H.M. (2001) Tratamento de odores em sistemas de esgotos sanitários. In *Pós-tratamento de efluentes de reatores anaeróbios* (C.A.L. Chernicharo coordenador) cap. 8, FINEP/PROSAB, Rio de Janeiro, Brasil, 544 p. (in Portuguese).

Campos, C.M.M. (1990) Physical aspects affecting granulation in UASB reactors. Ph.D. thesis, University of Newcastle upon Tyne, UK.

Campos, J.R. (1994) Alternativas para tratamento de esgotos – Pré-tratamento de águas para abastecimento. Consórcio Intermunicipal das Bacias dos Rios Piracicaba e Capivari, no. 9, 112 p. (in Portuguese).

Campos, C.M.M. and Chernicharo, C.A.L (1990) The use of the SMA-test for measuring toxicity in anaerobic sludge. In *Proc. IAWPRC International Seminar on Industrial Residuals Management*, Salvador, Brazil, pp. 191–199.

Campos, J.R. and Dias, H.G. (1989) Potencialidades do filtro anaeróbio. *Revista DAE* **49** n.154 (in Portuguese).

Campos, J.R. and Pagliuso, J.D. (1999) Tratamento de gases gerados em reatores anaeróbios. In *Tratamento de esgotos sanitários por processo anaeróbio e disposição controlada no solo*, (J.R. Campos coordenador), cap. 10, FINEP/PROSAB, Rio de Janeiro, Brasil, 435 p. (in Portuguese).

Capri, M.G. and Marais, G.V.R. (1975) pH adjustment in anaerobic digestion. *Water Res.* **9**, 307–313.

Cassini, S.T., Chernicharo, C.A.L., Andreoli, C.V., França, M., Borges E.S.M and Gonçalves, R.F. (2003) Hidrólise e atividade anaeróbia em lodos. In *Digestão anaeróbia de resíduos sólidos orgânicos e aproveitamento de biogás*, (S.T. Cassini coordenador), cap. 2, FINEP/PROSAB, Rio de Janeiro, Brasil, 196 p. (in Portuguese).

Cavalcanti, P.F.F, van Haandel, A.C., Kato, M.T., von Sperling, M., Luduvice, M.L and Monteggia, L.O. (2001) Pós-tratamento de efluentes de reatores anaeróbios por lagoas de polimento. In *Pós-tratamento de efluentes de reatores anaeróbios*, (C.A.L. Chernicharo coordenador), cap. 3, FINEP/PROSAB, Rio de Janeiro, Brasil, 544 p. (in Portuguese).

Chernicharo, C.A.L. (1990) The effect of temperature and substrate concentration on the performance of U.A.S.B. reactors. Ph.D. thesis, University of Newcastle upon Tyne, UK.

Chernicharo, C.A.L. (1997) *Princípios do tratamento biológico de águas residuárias – Volume 5: Reatores anaeróbios.* Departamento de Engenharia Sanitária e Ambiental, DESA/UFMG, Belo Horizonte, Brasil, 245 p.

Chernicharo, C.A.L. and Campos, C.M.M. (1990) A new methodology to evaluate the behaviour of anaerobic sludge exposed to potentially inhibitory compounds. In *Proc. IAWPRC International Seminar on Industrial Residuals Management*, Salvador, Brazil, pp. 79–88.

Chernicharo, C.A.L. and Campos, C.M.M. (1991) O uso do respirômetro *Warburg* para a avaliação da atividade metanogênica específica de lodos anaeróbios. In *Anais do 16ᵃ Congresso Brasileiro de Engenharia Sanitária e Ambiental*, vol. 2, Tomo I, 272–285 (in Portuguese).

Chernicharo, C.A.L. and Campos, C.M.M. (1995) *Tratamento anaeróbio de esgotos.* Depto. de Engenharia Sanitária e Ambiental da Escola de Engenharia da UFMG, Apostila, 65 p. (in Portuguese).

Chernicharo, C.A.L., van Haandel, A.C. and Cavalcanti, P.F.F. (1999) Capítulo 9: Controle operacional de reatores anaeróbios. In *Tratamento de esgoto sanitário por processo anaeróbio e disposição controlada no solo*, (J.R. Campos coordenador), ABES/FINEP/PROSAB, Rio de Janeiro, Brasil, 436 p. (in Portuguese).

Chernicharo, C.A.L., Cota, R.S., Zerbini, A.M., von Sperling, M. and Castro. Brito, L.N. (2001a) Post-treatment of anaerobic effluents in an overland flow system. *Water Sci. Technol.* **44** (4), 229–236.

Chernicharo, C.A.L., van Haandel, A.C., Cybis, L.F. and Foresti, E. (2001b) Post treatment of anerobic effluents in Brazil: state of the art. In *Proceedings of the 9ᵗʰ World Congress on Anaerobic Digestion.* Technologisch Instituut, IWA, Netherlands Association for Water Management. Antuerp, Belgium, pp. 747–752.

Chernicharo, C.A.L., van Haandel, A.C., Foresti, E. and Cybis, L.F. (2001c) Introdução. In *Pós-tratamento de efluentes de reatores anaeróbios*, (C.A.L. Chernicharo, coordenador), cap. 1, FINEP/PROSAB, Rio de Janeiro, Brasil, 544 p. (in Portuguese).

Christensen, D.R., Gerick, J.A. and Eblen, J.E. (1984) Design and operation of an upflow anaerobic sludge blanket reactor. *J. Water Pollut. Control Fed.* **56**, 1059–1062.

Cohen, A. (1991) Effects of some industrial chemicals on methanogenic activity measured by sequential automated methanometry (SAM). In *Proc. VI International Symposium on Anaerobic Digestion*, São Paulo, Brazil, 11–20.

Coraucci Filho, B., Andrade Neto, C.O, Melo, H.N.S., Souza J.T., Nour, E.A.A. and Figueiredo, R.F. (2001) Pós-tratamento de efluentes de reatores anaeróbios por sistemas de aplicação no solo. In *Pós-tratamento de efluentes de reatores anaeróbios*, (C.A.L. Chernicharo coordenador), FINEP/PROSAB, Rio de Janeiro, Brasil, 544 p. (in Portuguese).

de Zeeuw, W. (1984) Acclimatization of anaerobic sludge for UASB-reactor start-up. Ph.D. thesis, Agricultural University, Wageningen, The Netherlands.

Dolfing, J. and Mulder, J.W. (1985) Comparison of methane production rate and coenzyme F_{420} content of methanogenic consortia in anaerobic granular sludge. *Appl. Environ. Microbiol.* **49**, 1142–1145.

Feachem, R.G. *et al.* (1983) Sanitation and disease. Health aspects of excreta and wastewater management. *World Bank Studies in Water Supply and Sanitation 3*, The World Bank.

Foresti E. (1994) Fundamentos do Processo de digestão anaerobia. In *Anais III Taller y Seminario Latinoamericano: tratamiento anaerobio de aguas residuales*, Montevideo, Uruguay, pp. 97–110 (in Portuguese).

Friedman, A.A. and Tait S.J. (1980) Anaerobic rotating biological contactor for carbonaceous wastewater. *J. Water Pollut. Control Fed.* **8**, 2257–2269.

Gaudy, A.F. Jr. and Gaudy E.T. (1980) *Microbiology for Environmental Scientists and Engineers*, McGraw-Hill, New York.

Ghosh, S., Conrad, J.R. and Klass, D.C. (1975) Anaerobic acidogenesis of wastewater sludge. *J. Water Pollut. Control Fed.* **47**, 30–47.

Gonçalves, R.F., Chernicharo, C.A.L., Andrade Neto, C.O, Alem Sobrinho, P., Kato, M.T., Costa, R.H.R., Aisse M.M., Zaiat, M. (2001). Pós-tratamento de efluentes de reatores anaeróbios por reatores com biofilme. Cap. 4. In: CHERNICHARO C.A.L. (coordenador). Pós-tratamento de efluentes de reatores anaeróbios. FINEP/PROSAB, Rio de Janeiro, Brasil, 544 p. (in Portuguese).

Gorris, L.G., de Kok, T.M., Kroon, B.M., van der Drift, C. and Vogels, G.D. (1988) Relationship between methanogenic cofactor content and maximum specific methanogenic activity of anaerobic granular sludges. *Appl. Environ. Microbiol.* **54**, 1126–1130.

Goyal, S.M. and Gerba, C.P. (1979) Comparative absorption of human enteroviruses, simian rotavirus, and selected bacteriophages to soils. *Appl. Environ. Microbiol.* **38**, 241–247.

Grady, C.P.L. and Lim, H. (1980) *Biological Wastewater Treatment: Theory and Application.* Marcel Dekker, New York, 963 p.

GTZ/TBW (1997). Promotion of anaerobic technology for the treatment of municipal and industrial sewage and wastes. *Supraregional sectoral project*

Guiot, S.R., Pauss, A. and Costerton, J.W. (1992) A structured model of the anaerobic granule consortium. *Water Sci. Technol.* **25** (7), 1–10.

Hamoda, M.F. and van den Berg, L. (1984) Effect of settling on performance of the upflow anaerobic sludge bed reactors. *Water Res.* **18**, 1561–1567.

Henze, M. and Harremöes, P. (1983) Anaerobic treatment of wastewater in fixed film reactors: a literature review. *Water Sci. Technol.* **15** (8/9), 1–101.

Hulshoff Pol, L.W. (1995) Waste characteristics & factors affecting reactor performance 1: Process factors, waste characterisation and pH related toxicity. In *International course on anaerobic treatment.* Wageningen Agricultural University/IHE Delft. Wageningen, 17–28 Jul. 1995.

Hulshoff Pol L.W., Dolfing J., van Strate, K., de Zeeuw, W.J. and Lettinga, G. (1984) Pelletization of anaerobic sludge in upflow anaerobic sludge bed reactors on sucrose-containing substrate. In *Proc. 3rd International Symposium on Microbial Ecology, American Society of Microbiology*, pp. 636–642.

Hungate, R.E. (1969) A rool-tube method for cultivation of strict anaerobes. In *Methods in Microbiology* (ed. J.R. Norris and D.W. Ribbons), chapter IV, Academic Press, London, pp. 117–132.

James, A., Chernicharo, C.A.L. and Campos, C.M.M. (1990) The development of a new methodology for the assessment of specific methanogenic activity. *Water Res.* **24**, 813–825.

Jewell, W.J. (1981) Development of the attached microbial film expanded bed process for aerobic and anaerobic waste treatment. In *Biological Fluidised Bed Treatment of Water and Wastewater* (eds P.F. Cooper and B. Atkinson) Ellis Horwood, Chichester, p. 251.

Kato, M.T. (1994) The anaerobic treatment of low strength soluble wastewaters. Ph.D. thesis. Agricultural University of Wageningen, The Netherlands.

Khalid, R.A. *et al.* (1978) Nitrogen removal processes in an overland flow treatment of wastewater. In *State of Knowledge in Land Treatment of Wastewater*, vol. 2, U.S. Army Corps of Engineers, Hanover, NH.

Lee, C.R. *et al.* (1976). Highlights of research on overland flow for advanced treatment of wastewater. Misc. Paper Y-76-6, U.S. Army Corps of Engineers, Waterways Experimental Station, Vicksburg, MS.

Lettinga, G. (1995a) Anaerobic reactor technology: reactor and process design. In *International Course on Anaerobic Treatment*. Wageningen Agricultural University/IHE Delft, Wageningen, 17–28 Jul. 1995.

Lettinga, G. (1995b) Introduction. In *International Course on Anaerobic Treatment*. Wageningen Agricultural University/IHE Delft, Wageningen, 17–28 Jul. 1995.

Lettinga, G., Hulshoff, Pol, L.W., Koster, I.W., Wiegant, W.M., de Zeeuw, W.J., Rinzema, A., Grin, P.C., Roersma, R.E. and Hobma, S.W. (1984) High-rate anaerobic wastewater treatment using the UASB reactor under a wide range of temperature conditions. *Biotechnol. Genetic Eng. Rev.* **2**, 253–284.

Lettinga, G., Hulshoff, Pol, L.W. and Zeeman, G. (1996) *Biological Wastewater Treatment. Part I: Anaerobic Wastewater Treatment*. Lecture Notes. Wageningen Agricultural University, edn January 1996.

Lettinga, G., Roersma, R. and Grin, P. (1983) Anaerobic treatment of raw domestic sewage at ambient temperatures using a granular bed UASB reactor. *Biotechnol. Bioeng.* **25**, 1701–1723.

Lettinga, G., van Velsen, A.F.M, de Zeeuw, W. and Hobma, S.W. (1979) Feasibility of the upflow anaerobic sludge blanket (UASB) process. In *Proc. Nat. Conf. Environmental Engineering*, ASCE, Virginia, Reston, pp. 35–45.

Lettinga, G., van Velsen, A.F.M., Hobma, S.W., de Zeeuw, W. and Klapwijc, A. (1980) Use of upflow sludge blanket (USB) reactor concept for biological wastewater treatment, especially for anaerobic treatment. *Biotechnol. Bioeng.* **22**, 699–734.

Malina, J.F. Jr. and Pohland, F.G. (1992) *Design of Anaerobic Processes for the Treatment of Industrial and Municipal Wastes*, vol. **7**, Technomic Publishing, Inc., USA.

Martel, C.J. *et al.* (1980) Rational design of overland flow systems. *Procs of the National Environmental Engineering Conference* (ed., W. Saukin), ASCE, New York, pp. 114.

McCarty, P.L. (1964) Anaerobic waste treatment fundamentals. *Public Works* – Parts 1, 2, 3 and 4. **95**, No. 9, pp. 107–112; No. 10, pp. 123–126; No. 11, pp. 91–94; No. 12, pp. 95–99.

McCarty, P.L. (1966) Kinetics of waste assimilation in anaerobic treatment. *American Institute of Biology Science Developments in Industrial Microbiology* **7**, 144–155.

McCarty, P.L. and McKinney, R.E. (1961) Volatile acid toxicity in anaerobic digestion. Salt toxicity an anaerobic digestion. *J. Water Pollut. Control Fed.* **33**, 399.

Metcalf and Eddy (1991) *Wastewater Engineering: Treatment, Disposal, and Reuse*, 3rd edn, Metcalf and Eddy, Inc. 1334 p.

Monteggia, L.O. (1991) The use of specific methanogenic activity for controlling anaerobic reactors. Ph.D. thesis, University of Newcastle upon Tyne, UK.

Novaes, R.F.V. (1986) Microbiology of anaerobic digestion. In *International Seminar Anaerobic Treatment in Tropical Countries*, Sao Paulo, Brazil.

Orenland, R.S. and Polcin, S. (1982) Methanogenesis and sulphate reduction: competitive and non-competitive substrates in estuarine sediments. *Appl. Environ. Microbiol.* **44**, 1270–1276.

Paganini, W.S. (1997) *Disposição de esgoto no solo (escoamento à superfície)*, Fundo editorial da AESABESP, São Paulo, 232 p. (in Portuguese).

Pena, J.A. (1994) Estudo da metodologia do teste de atividade metanogênica específica. *Tese de Doutorado*. Escola de Engenharia de São Carlos – USP (in Portuguese).

Pessôa, C.A. and Jordão, E.P. (1982) *Tratamento de esgotos domésticos*. Convênio ABES/BNH. (in Portuguese).

Peters, R.E. and Lee, C.R. (1978) Field investigations of advanced treatment of municipal wastewater overland flow. In *State of Knowledge in Land Treatment of Wastewater*, vol. 2, U.S. Army Corps of Engineers, Hanover, NH, p. 45.

Pinto, J.D.S. (1995) Tratamento de esgotos domésticos através de filtro anaeróbio utilizando escória de alto forno como meio suporte. *Dissertação de mestrado*. Departamento de Engenharia Sanitária e Ambiental da UFMG (in Portuguese).

Ripley, L.E., Boyle, W.C. and Converse, J.C. (1983) Improved alkalimetric monitoring for anaerobic digestion of high-strength wastes. *J. WPCF* **58** (5), 406–411.

Schaub, S.A. *et al.* (1978) Evaluation of the overland runoff mode of land wastewater application for virus removal. In *State of Knowledge in Land Treatment of Wastewater*, vol. 2, U.S. Army Corps of Engineers, Hanover, NH, p. 245.

Schimidt, J.E. and Ahring, B.K. (1996) Granular sludge formation in upflow anaerobic sludge blanket (UASB) reactors. Biotech. Bioeng. 4, 229–246.

Soubes, M. (1994) Microbiologia de la digestion anaerobia. In *Anais III Taller y Seminario Latinoamericano: tratamiento anaerobio de aguas residuales*. Montevideo, Uruguay, pp. 15–28.

Speece, R.E. (1983) Environmental requirements for anaerobic digestion of biomass. In *Advances in Solar Energy – An Annual Review of Research and Development*.

Speece, R.E. and McCarty, P.L. (1964) Nutrient requirements and biological solids accumulation in anaerobic digestion. *Adv. Water Pollut. Res.*, vol. **2**, Pergamon Press, London, 305 p.

Speece, R.E. *et al.* (1986) *Proc. EWPCA Conf. Anaerobic Treatment, a Grown-up Technology*, September 1986, Amsterdam, 205 p.

Stainer, R.Y., Adelberg, E.A. and Indraham, J.L. (1976) *The Microbial World*, 4th edn, Prentice-Hall, Inc., Englewood Cliffs, New Jersey.

Stien, J.L. and Schwartzbrod, J. (1990) Experimental contamination of vegetables with helminth eggs. Water *Sci. Technol.* **22** (9), 51–57.

Stronach, S.M., Rudd, T. and Lester, J.N. (1986) *Anaerobic Digestion Processes in Industrial Wastewater Treatment*, Springer-Verlag, Berlin.

Switzenbaum, M.S. and Jewell, W.J. (1980) Anaerobic attached-film expanded-bed reactor treatment. *J. Water Pollut. Control Fed.* **52**, 1953–1965.

Umbreit, W.W., Burris, R.H. and Stayffer, J.F. (1964) *Manometric Techniques, 4th edn.*, Burgess, Berks.

U.S. Environmental Protection Agency (1981) *Process Design Manual: Land Treatment of Municipal Wastewater*. United States Environmental Protection Agency, Cincinnati, Ohio.

U.S. Environmental Protection Agency (1982) *Chemicals in your Community, A Guide to the Emergency Planning and Community Right-to-Know Act*. EPA-516/002-80246, Washington, D.C.

U.S. Environmental Protection Agency (1986) *Municipal Wastewater Disinfection – Design Manual, EPA/625/1-86/021*, Cincinnati, 247 p.

van den Berg, L. and Kennedy, K.J. (1981) Support materials for stationary fixed film reactors for high-rate methanogenic fermentation. *Biotechnol. Lett.* **3**, 165–170.

van den Berg, L. and Lentz, C.P. (1977) Food processing waste treatment by anaerobic digestion. *Proc. of the 32nd Industrial Waste Conference*, Purdue University, Lafayette, Indiana, pp. 252–258.

van Haandel, A.C. and Lettinga, G. (1994) *Anaerobic Sewage Treatment: A Practical Guide for Regions with a Hot Climate*, John Wiley and Sons, 222 pp.

van Lier, J.B., Rebac, S. and Lettinga, G. (1997) High rate anaerobic wastewater treatment under psychrophilic and thermophilic conditions. *Water Sci. Technol.*, **35**, 199–206.

van Lier, J.B., Zeeman and Huibers, F. (2002) Anaerobic (pre-) treatment for the decentralised reclamation of domestic wastewater, stimulating agricultural reuse. In *Proceedings VII Latin American Workshop and Symposium on Anaerobic Digestion*, Mérida, Yucatán, Mexico, pp. 534–541.

Visser, A. (1995) Anaerobic treatment of sulphate containing waste water. In *International Course on Anaerobic Treatment*. Wageningen Agricultural University/IHE Delft. Wageningen, 17–28 Jul. 1995.

von Sperling, M. (1995) *Introdução à qualidade das águas e ao tratamento de esgotos.* Departamento de Engenharia Sanitária e Ambiental – UFMG, Belo Horizonte, 240 p. (in Portuguese).

von Sperling, M. (1996) *Lagoas de estabilização.* Departamento de Engenharia Sanitária e Ambiental – UFMG. Belo Horizonte. 134 p. (in Portuguese).

von Sperling, M. and Chernicharo, C.A.L. (1996) *Sectorial Pilot Project on Anaerobic Digestion of Waste Water and Solid Waste.* GTZ/TBW: Interim report.

Water Environment Federation (1992) Design of municipal wastewater treatment plants. *Manual of Practice n. 8*, Alexandria, VA. ASCE Manuals and Reports on Engineering Practice no. 76, New York, NY.

Water Environment Federation (1996) *Operation of Municipal Wastewater Treatment Plants*, v. 1 Alexandria, Virginia, 474 p.

Water Pollution Control Federation (1983) *Nutrient Control.* Manual of Practice FD-7, Alexandria, Virginia, 205 pp.

Water Pollution Control Federation (1990) *Natural Systems for Wastewater Treatment.* Manual of Practice FD-16, Alexandria, Virginia, 270

WEF and ASCE (1992) *Design of Municipal Wastewater Treatment Plants*, Water Environment Federation/American Society of Civil Engineers, 1592 p.

Weiland, P. and Rozzi, A. (1991) The start-up, operation and monitoring of high-rate anaerobic treatment systems: discusser's report. *Water Sci. Technol.* **24**, 257–277.

Wiegant, W.M. and Lettinga, G. (1985) Thermophilic anaerobic digestion of sugars in upflow anaerobic sludge blanket reactors. *Biotechnol. Bioeng.* **27**, 1602–1607.

Winfrey, M.R. and Zeikus, J.G. (1977) Effect of sulphate on carbon and electron flow during microbial methanogenesis in freshwater sediments. *Appl. Environ. Microbiol.* **33**, 275–281.

World Health Organisation (1985) *Health Aspects of Nightsoil and Sludge Use in Agriculture and Aquaculture. IRCWD News*, WHO International Reference Centre for Wastes Disposal, Duebendorf, Switzerland.

World Health Organisation (1989) Health guidelines for use of wastewater in agriculture and acquaculture. *Technical Report Series*, vol. **778**, WHO, Geneva.

Young, J.C. (1991) Factors affecting the design and performance of upflow anaerobic filters. *Water Sci. Technol.* **24**, 133–155.

Young, J.C. and McCarty, P.L. (1969) Anaerobic filter for waste treatment. *J. Water Pollut. Control Fed.*, **41**, R160–R173.

Yspeert, P., Vereijken, T. Vellinga, S. and de Vegt, A. (1995) The IC reactor for anaerobic treatment of industrial wastewater. In *International Course on Anaerobic Treatment*. Wageningen Agricultural University/IHE Delft, Wageningen, 17–28 Jul. 1995.

Zehnder, A.J.B. (1978) Ecology of methane formation. In *Microbiology of Water Pollution*, Wiley Interscience Pub., New York, pp. 349–376.

Zehnder, A.J.B., Ingvorsen K., and Marti T. (1982) Microbiology of methane bacteria. *Anaerobic Digestion 1981*. Elsevier Biomedical Press, Amsterdam, The Netherlands.

Zerbini, A.M. and Chernicharo, C.A.L. (2001) Metodologias para quantificação, identificação e análise de viabilidade de ovos de helmintos em esgotos brutos e tratados – p. 70 – 107. In *Pós-tratamento de efluentes de reatores anaeróbios – aspectos metodológicos*, (C.A.L. Chernicharo coordenador), FINEP/PROSAB, Rio de Janeiro, Brasil, 107 p. (in Portuguese).

Zinder, S.H. (1993) Physiological ecology of methanogens. In *Methanogenesis, Ecology, Physiology, Biochemistry and Genetics* (ed. J.G. Ferry), Chapman and Hall, New York.

IWA Publishing's authorised EU representative for General Product
Safety Regulations is Diane D'Arras, 15 rue Duret, 75116 Paris,
France, e-mail: safety@iwap.co.uk.

Printed and bound by CPI Group (UK) Ltd, Croydon, CR0 4YY
12/05/2026
02108866-0001